T0181458

Lecture Notes in Computer Science　　9548

Commenced Publication in 1973
Founding and Former Series Editors:
Gerhard Goos, Juris Hartmanis, and Jan van Leeuwen

More information about this series at http://www.springer.com/series/7408

Jan Kofroň · Tomáš Vojnar (Eds.)

Mathematical and Engineering Methods in Computer Science

10th International Doctoral Workshop, MEMICS 2015
Telč, Czech Republic, October 23–25, 2015
Revised Selected Papers

 Springer

Editors
Jan Kofroň
Charles University
Prague
Czech Republic

Tomáš Vojnar
Brno University of Technology
Brno
Czech Republic

ISSN 0302-9743   ISSN 1611-3349   (electronic)
Lecture Notes in Computer Science
ISBN 978-3-319-29816-0   ISBN 978-3-319-29817-7   (eBook)
DOI 10.1007/978-3-319-29817-7

Library of Congress Control Number: 2016930670

LNCS Sublibrary: SL2 – Programming and Software Engineering

This Springer imprint is published by SpringerNature
The registered company is Springer International Publishing AG Switzerland

Preface

This volume contains the proceedings of the 10th Doctoral Workshop on Mathematical and Engineering Methods in Computer Science (MEMICS 2015) held in Telč, Czech Republic, during October 23–25, 2015.

The aim of the MEMICS workshop series is to provide an opportunity for PhD students to present and discuss their work in an international environment. The scope of MEMICS is broad and covers many fields of computer science and engineering. In 2015, we paid special attention to submissions in the following (although not exclusive) areas:

- Security and safety
- Bioinformatics
- Recommender systems
- High-performance and cloud computing
- Non-traditional computational models (quantum computing, etc.)

There were 25 submissions from PhD students from nine countries. Each submission was thoroughly evaluated by at least three Program Committee members, who also provided extensive feedback to the authors. Out of these submissions, ten full papers were selected for publication in these proceedings, and additional seven papers were selected for presentation at the workshop.

The highlights of the MEMICS 2015 program included six keynote lectures delivered by internationally recognized researchers from the aforementioned areas of interest. The speakers were:

Ezio Bartocci (Vienna University of Technology, Austria)
Siegried Benkner (University of Vienna, Austria)
Mike Just (Heriot-Watt University, Edinburgh, UK)
Simone Severini (University College London, UK)
Natasha Sharygina (University of Lugano, Switzerland)
Peter Vojtáš (Charles University, Czech Republic)

The full papers of three of these keynote lectures are also included in the proceedings. In addition to regular papers, MEMICS workshops traditionally invite PhD students to submit a presentation of their recent research results that have already undergone a rigorous peer-review process and have been presented at a high-quality international conference or published in a recognized journal. A total of 12 presentations out of 14 submissions from 11 countries were included into the MEMICS 2015 program.

The MEMICS tradition of best paper awards continued also in the year 2015. The best contributed papers were selected during the workshop, taking into account their scientific and technical contribution together with the quality of presentation. The 2015 awards went to the following papers:

Vojtěch Havlena and Dana Hliněná: "Fitting Aggregation Operators"

Agnis Arins: "Span-Program-Based Quantum Algorithms for Graph Bipartiteness and Connectivity"

The two awards consisted of a diploma accompanied by a financial prize of 400 Euro each. The prize money was donated by ZONER software, a. s. and Brno University of Technology.

The successful organization of MEMICS 2015 would not have been possible without generous help and support from the organizing institutions: Brno University of Technology, Masaryk University in Brno, and Charles University Prague.

We thank the Program Committee members and the external reviewers for their careful and constructive work. We thank the Organizing Committee members who helped to create a unique and relaxed atmosphere that distinguishes MEMICS from other computer science meetings. We also gratefully acknowledge the support of the EasyChair system and the great cooperation with the Lecture Notes in Computer Science team of Springer.

November 2015   Jan Kofroň
                                                                Tomáš Vojnar

Organization

The 10th Doctoral Workshop on Mathematical and Engineering Methods in Computer Science (MEMICS 2015) took place in Telč, Czech Republic, during October 23–25, 2015. More information about the MEMICS workshop series is available at http://www.memics.cz.

General Chair

Tomáš Vojnar   Brno University of Technology, Brno, Czech Republic

Program Committee Chair

Jan Kofroň   Charles University Prague, Czech Republic

Program Committee

Andris Ambainis	University of Latvia, Latvia
Gianni Antichi	University of Cambridge, UK
Michal Baczynski	Uniwersytet Ślaski w Katowicach, Poland
Ezio Bartocci	Vienna University of Technology, Austria
Jan Bouda	Masaryk University, Czech Republic
Krishnendu Chatterjee	IST Austria, Austria
Markus Chimani	Osnabrück University, Germany
Pavel Čeleda	Masaryk University, Czech Republic
Eva Dokládalová	ESIEE Paris, France
Martin Drahanský	Brno University of Technology, Czech Republic
Fréderic Dupuis	Masaryk University, Czech Republic
Jiří Filipovič	Masaryk University, Czech Republic
Vojtěch Forejt	University of Oxford, UK
Dieter Gollmann	TU Hamburg, Germany
Derek Groen	University College London, UK
Petr Hanáček	Brno University of Technology, Czech Republic
Lucia Happe	KIT, Germany
Marcus Huber	Universitat Autónoma de Barcelona, Spain
Antti Hyvärinen	University of Lugano, Switzerland
Ondřej Jakl	VŠB-TU Ostrava, Czech Republic
Jiří Jaroš	Brno University of Technology, Czech Republic
Stefan Kiefer	University of Oxford, UK
Lukasz Kowalik	University of Warsaw, Poland
Stanislav Krajci	Pavol Jozef Šafárik University in Košice, Slovak Republic
Dieter Kranzlmüller	Ludwig Maxmillian University and Supercomputing Centre Garching Munich, Germany

Daniel Král'	University of Warwick, UK
Daniel Langr	Czech Technical University in Prague, Czech Republic
Luděk Matyska	Masaryk University, Czech Republic
Andrzej Mizera	University of Luxembourg, Luxembourg
Manuel Ojeda Aciego	University of Malaga, Spain
Marcin Pawlowski	University of Gdansk, Poland
Cristina Seceleanu	Mälardalen University, Sweden
Jiří Srba	Aalborg University, Denmark
Ivan Šimeček	Czech Technical University in Prague, Czech Republic
Josef Šlapal	Brno University of Technology, Czech Republic
Catia Trubiani	Gran Sasso Science Institute, Italy
Thomas Vetterlein	Johannes Kepler University, Austria
Mário Ziman	Slovak Academy of Sciences, Slovak Republic
Florian Zuleger	TU Wien, Austria

Steering Committee

Tomáš Vojnar, Chair	Brno University of Technology, Brno, Czech Republic
Milan Češka	Brno University of Technology, Brno, Czech Republic
Zdeněk Kotásek	Brno University of Technology, Brno, Czech Republic
Mojmír Křetínský	Masaryk University, Brno, Czech Republic
Antonín Kučera	Masaryk University, Brno, Czech Republic
Luděk Matyska	Masaryk University, Brno, Czech Republic

Organizing Committee

Radek Kočí, Chair	Brno University of Technology, Czech Republic
Tomáš Fiedor	Brno University of Technology, Czech Republic
Hana Pluháčková	Brno University of Technology, Czech Republic
Jaroslav Rozman	Brno University of Technology, Czech Republic
Lenka Turoňová	Brno University of Technology, Czech Republic

Additional Reviewers

Stanislav Böhm
Robin Kothari
Tobias Fuchs

Contents

Programming Support for Future Parallel Architectures

Siegfried Benkner[✉]

Research Group Scientific Computing, University of Vienna, Vienna, Austria
siegfried.benker@univie.ac.at

Abstract. Due to physical constraints the performance of single processors has reached its limits, and all major hardware vendors switched to multi-core architectures. In addition, there is a trend towards heterogeneous parallel systems comprised of conventional multi-core CPUs, GPUs, and other types of accelerators. As a consequence, the development of applications that can exploit the potential of emerging parallel architectures and at the same time are portable between different types of systems is becoming more and more challenging. In this paper we discuss recent research efforts of the European PEPPHER project in software development for future parallel architectures. We present a high-level compositional approach to parallel software development in concert with an intelligent task-based runtime system. Such an approach can significantly enhance programmability of future parallel systems, while ensuring efficiency and facilitating performance portability across a range of different architectures.

1   Introduction

Computer architectures are currently undergoing a significant shift from homogeneous multi-core designs to heterogeneous many-core systems, which combine different types of execution units like conventional CPU cores, GPUs and other accelerators within a single chip or compute node. While heterogeneous many-core architectures promise to deliver superior performance and energy efficiency, these architectures sharply increase the complexity of software development. Ensuring both a reasonable level of performance and a sufficient degree of performance portability of software between different systems is a fundamental challenge for current computer science research and engineering. In general, there is no guarantee that software developed for a particular architecture will be executable on another, related architecture. Even if functional portability is achieved, there is no guarantee that the performance achieved on a specific architecture will be preserved to a similar extent on other architectures. Expensive manual work in adapting, optimizing or rewriting an application for a specific architecture may thus be lost when porting the application to another, next generation architecture. There is therefore an urgent need for techniques

© Springer International Publishing Switzerland 2016
J. Kofroň and T. Vojnar (Eds.): MEMICS 2015, LNCS 9548, pp. 1–10, 2016.
DOI: 10.1007/978-3-319-29817-7_1

that facilitate efficient, productive and portable programming of heterogeneous many-core systems, including means for preserving aspects of performance when porting applications across different architectures.

The European PEPPHER project [4, 10] addressed programmability and performance portability for single-node heterogeneous many-core architectures, typically systems comprising multi-core CPUs with one or more GPUs, coprocessors or other types of accelerators. Key to the PEPPHER approach is a performance-aware component model in concert with a multi-level parallel task-based execution model. Within this model, programs are composed at a high-level of abstraction from sequential or already parallelized program components. For each component, different implementation variants optimized for different types of cores are provided by skilled expert programmers or taken from vendor-supplied libraries. A component-based high-level program is transformed into code which employs the StarPU [2] runtime system to select suitable component implementation variants and to schedule their parallel execution in a performance- and resource-aware manner to the different execution units of a heterogeneous many-core architecture, exploiting all levels of parallelism provided by the hardware.

In this paper we focus on describing some major aspects of the PEPPHER framework. Other developments, including a toolbox with autotuned data structures and algorithms, hardware support for programmability and performance portability, the software simulator for experimentation with future heterogeneous many-core designs, and the PEPPHER Processing Unit (PePU), an experimental heterogeneous many-core hardware platform, are not covered.

2   The PEPPHER Methodology and Framework

The central aspects of PEPPHER for improving programmability and performance portability are (1) to provide performance-critical parts of applications as components with multiple implementation variants tailored for different types of execution units in a heterogeneous many-core system and (2) to employ a performance-aware runtime scheduling strategy that dynamically determines which implementation variants to execute on which execution units of the target system such that performance objectives are optimized. By decoupling the specification of component functionality from the actual implementation, the runtime system can dynamically adapt applications to different hardware configurations of a target platform without requiring source code changes.

Figure 1 illustrates the PEPPHER methodology, where mainstream programmers construct applications from components at a high level of abstraction, expert programmers provide component implementation variants optimized for different execution units available in a heterogeneous many-core system, and the runtime system selects component implementation variants and schedules them for execution on the target architecture with the goal of minimizing overall execution time and utilizing all available execution units as efficiently as possible.

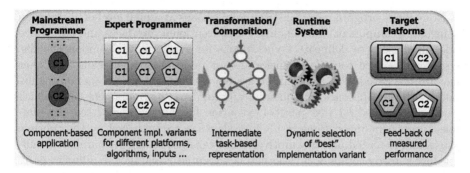

Mainstream Programmer	Expert Programmer	Transformation/ Composition	Runtime System	Target Platforms
Component-based application	Component impl. variants for different platforms, algorithms, inputs ...	Intermediate task-based representation	Dynamic selection of "best" implementation variant	Feed-back of measured performance

Fig. 1. The PEPPHER methodology of program development for heterogeneous many-core systems. C/C++ applications are annotated by the mainstream programmer to delineate the use of components. The more skilled, expert programmer provides component implementation variants optimized for specific hardware. A transformation or composition tool generates the necessary glue code such that component execution is delegated to the runtime system. At runtime a PEPPHER program is represented by a dynamically constructed acyclic task graph where each node (task) corresponds to a component invocation and edges indicate data dependences. The task graph is processed by the runtime system which selects proper implementation variants for tasks and schedules them for execution on the available execution units of the target system.

A PEPPHER component is a self-contained, side-effect free functional unit that implements a specific functionality declared in an interface in several different implementation variants, typically variants for each type of execution unit (CPU, GPUs, etc.) available in a heterogeneous target architecture. Component implementation variants are usually written by expert programmers possibly using different programming APIs (e.g., CUDA, OpenCL) or are taken from optimized vendor-supplied libraries. Further specialized variants may be generated, e.g., by means of autotuning. Components and implementation variants are accompanied with rich meta-data, supplied via external XML descriptors. Besides information about the data read and written by components, meta-data includes information about resource requirements, possible target platforms, and performance relevant parameters. For this purpose a Platform Description Language (PDL) [16] for describing essential hardware- and software characteristics of heterogeneous many-core systems has been developed. The PDL allows describing a hierarchical aggregation of system components comprising processing units, memory regions and interconnect capabilities. Component implementation variants may be associated with generic performance models for estimating at runtime their (relative) performance on different execution units. Performance models may rely on the PDL to determine capabilities of potential target execution units and might be realized using different approaches ranging from analytical models to regression-based performance estimation using historical performance data.

Non-expert programmers construct applications at a high level of abstraction by invoking component functionality via conventional interfaces and use source code annotations to delineate asynchronous (or synchronous) component calls. A source-to-source compiler transforms annotated component calls such that they are registered with the runtime system and generates corresponding glue-code. With this approach, a sequential program spawns component calls, which are then scheduled for task-parallel execution by the runtime system. At runtime, component invocations result in tasks that are managed by the runtime system and executed non-preemptively. The execution model is parallel at multiple levels: ready component tasks can be executed in parallel on different parts of the system, and component tasks can themselves be parallel, e.g., OpenCL or CUDA variants for the GPU and multi-core parallel variants for the CPU.

The PEPPHER framework and methodology makes it possible to gradually make an existing application more efficient for a given, heterogeneous parallel system, as well as more performance portable across different types of heterogeneous systems, by progressively supplying more suitable and efficient component variants, and by outlining more and more parts of the application into components.

Figure 2 illustrates the PEPPHER software stack to assist in development and generation of efficient, performance portable applications for heterogeneous many-core systems.

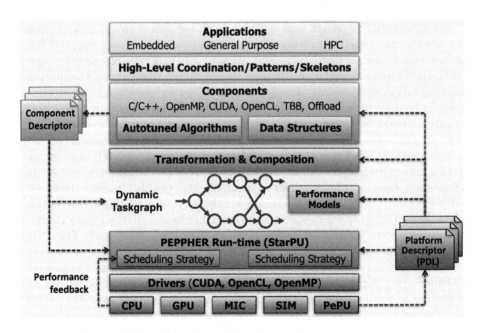

Fig. 2. The PEPPHER software stack.

3   Coordination Language and Transformation Tool

The PEPPHER coordination language provides a small set of directives (pragmas) for coordinating the use of PEPPHER components from within C/C++ applications at a high level of abstraction. The coordination language enables users to indicate that certain functionality of a program should be realized by means of components, while delegating to the runtime system the decision which implementation variants to select. By means of basic coordination directives, calls to components may be performed asynchronously or synchronously. Additional information that may be provided at component call sites includes parameter assertions, performance expectations, preferred execution targets, and data partitioning information and access patterns for array parameters.

```
#pragma pph call
cf1(A, N, B, M);       //A:read, B:write (→ external XML meta-data)

#pragma pph call sync
cf2(C, M);             //block until cf2 returns

#pragma pph call parameter(size < 1000) //assertion
cf3(D, size);

#pragma pph call target(OPENCL)        //run on some OpenCL device
cf4(E, N);
```

```
...
#pragma pph pipeline buffer(UNORDERED,?) //autotune buffer size
while(inputstream >> file) {
    readImage(file,image);
    #pragma pph stage replicate(?)    //autotune replication factor
    {
        resizeAndColorConvert(image);
        detectFace(image,outImage);
    }
    writeImage(file,outImage);
}
```

Fig. 3. Examples of PEPPHER coordination constructs.

Besides providing the user with means for integrating PEPPHER components into an application, the coordination language offers high-level support for the structured implementation of parallel patterns, in particular for pipelining [3], as well as for task farming [7]. Using all these features, PEPPHER supports multiple forms of parallelism, including task parallelism between asynchronous component invocations, data parallelism within components, as well as pipelining across components. Examples of basic component calls as well as a pipeline pattern are shown in Fig. 3.

Two prototype implementations of the PEPPHER framework have been developed, a transformation tool and a composition tool. In the following we outline some aspects of the transformation tool. More information about the composition tool can be found in [7].

The PEPPHER transformation tool is a source-to-source transformation system that transforms a C++ application with coordination language pragma directives into a C++ code with calls to the runtime system. When executed, the generated target code submits for each component call a task to the runtime system resulting in a dynamic, directed acyclic graph of component tasks with data dependencies. If a task has multiple implementation variants it is up to the runtime to chose the best variant such that overall performance is optimized.

While basic component calls are transformed into code that submits tasks directly to the underlying StarPU runtime system, pipeline patterns are transformed in such a way that they utilize a special pipeline coordination layer [3], which has been implemented on top of StarPU. The pipeline coordination layer controls all higher-level aspects of pipeline patterns, including the automatic generation and management of pipeline stages and the data transfer between stages by means of buffers. The pipeline coordination layer also controls the granularity of parallelism, e.g. using an external autotuner, by replicating compute intensive pipeline stages such that the execution is as balanced as possible and decides when and how many tasks are submitted to the runtime system. For each instance of a pipeline stage the coordination layer registers a task with the runtime system which is then responsible for executing the component associated with this stage. The transformation systems utilizes a component repository to look up component descriptors and a PDL descriptor of the target architecture to preselect component implementation variants to be considered for execution by the runtime system.

4   Task-Based Heterogeneous Runtime System

Execution of a PEPPHER application is managed by StarPU [2,8], a flexible, performance- and resource-aware, heterogeneous run-time system.

StarPU relies on a representation of the program as a directed acyclic graph (DAG) where nodes represent component calls (tasks) and edges represent data dependences. The runtime system dynamically schedules component calls to the available execution units of a heterogeneous many-core architecture such that (1) independent component calls execute in parallel on different execution units and (2) the "best" implementation variants for a given architecture are selected based on (historical) performance information captured in performance models.

Run-time component implementation variant selection is based on optimization objectives (e.g., minimizing execution time), resource availability, data availability and placement, and available performance information for the variants, while respecting data dependencies between components. Performance information, input information, optimization criteria, resource requirements and availability, data placement in the system, e.g., in main CPU or in GPU memory, are all used to determine which of the ready component task variants are scheduled onto which execution unit(s) of the system.

For CPU-GPU based systems with separate memory spaces, StarPU implements a software virtual shared memory with relaxed consistency and data

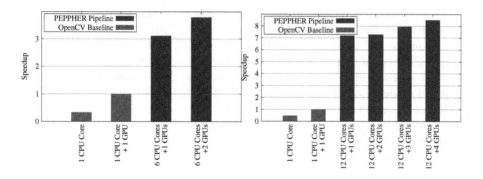

Fig. 4. Speedup results for a face detection application on various configurations of two different CPU/GPU systems relative to the OpenCV baseline version which can utilize only one CPU core and one GPU.

replication capability. The application just has to register the different pieces of data by providing their addresses and sizes in the main memory. To handle data distribution among processing units, StarPU also provides an additional high-level abstraction to easily handle partitioned data for block- and tile-based parallelism. Besides the Heterogeneous Earliest Finish Time (HEFT) scheduling policy [17], which is the default strategy, StarPU supports eager scheduling, priority scheduling and several variants of work-stealing. The HEFT policy considers inter-component data dependencies, and schedules components to workers taking into account the current system load, available component implementation variants, and historical execution profiles, with the goal of minimizing overall execution time by favoring implementations variants with the lowest expected execution time.

5 Experimental Results

Among several other benchmarks and applications, we have implemented a computer vision application for detecting faces in a stream of images based on pipeline patterns (see Fig. 3). For the detection stage two different component implementation variants, one for CPUs and one for GPUs, have been re-engineered from the popular Open Source Computer Vision (OpenCV) library [5] and utilized within the PEPPHER framework.

For performance comparison, we use a baseline OpenCV implementation, which, however, due to the restrictions of the OpenCV library, can only exploit a single GPU.

Figure 4 shows speedup results achieved with the PEPPHER implementation on different configurations of two different CPU/GPU systems. The first machine is equipped with two quad-core Intel Xeon X5550 CPUs and NVIDIA Tesla C2050 and C1060 GPUs, respectively. The second machine is equipped with two octa-core Intel Xeon E5-2650 CPUs and 4 NVIDIA Kepler K20 GPUs.

The left-hand side diagram in Fig. 4 shows speedup results for the Tesla-based GPU system relative to the baseline OpenCV version, which, as mentioned, can only utilize one CPU core and one GPU. The PEPPHER variant can harness multiple CPU cores and multiple GPUs without any source code changes and achieves a speedup of more than three over the OpenCV baseline. Compared to a sequential (single core) execution, the PEPPHER implementation achieves a speedup of about a factor of 20 using all available resources in the system. The right-hand side diagram in Fig. 4 shows the results for the Kepler-based GPU system. Again significant speedups are achieved by utilizing multiple CPU cores and GPUs. As opposed to the baseline OpenCV version, the PEPPHER version can seamlessly take advantage of all available computing resources of a heterogeneous system demonstrating that performance portability can be achieved without any changes to the high-level code.

Results of a performance evaluation of the PEPPHER composition tool with benchmarks from the RODINIA suite and other codes can be found in [7]. All these results indicate that with PEPPHER's high-level approach, the same source code can adapt to different target architecture configurations thanks to the flexibility offered by the framework and the dynamic scheduling facilities of the runtime system.

6   Related Work

Language, compiler and runtime support for future parallel architectures is a very active research area with many research efforts and projects world-wide. We can discuss only a few related approaches here. Several European projects addressed different aspects of programming support for multi- and many-core architectures including 2PARMA, APPLE-CORE, ENCORE, PARAPHRASE, RePhrase and others (see https://www.hipeac.net/network/projects/). In contrast to many of these efforts, PEPPHER is not focusing on providing a common programming model or virtual machine type portability layer to cope with heterogeneity. Through the use of component interfaces PEPPHER shields the mainstream programmer from architectural details while facilitating the use of vendor-specific libraries and enabling expert programmers to optimize implementation variants down to the metal using low-level APIs.

A key role in PEPPHER is played by the performance- and resource-ware StarPU runtime system, which dynamically adapts a program to the actual target platform by selecting suitable implementation variants. Other task-based runtime systems that share similarities with StarPU include OmpSs [6], HPX [9] and the Open Community Runtime OCR [14].

As opposed to implicit parallelization and performance portability via domain specific languages, as for example in [11], PEPPHER is taking a general-purpose approach, but also provides support for common parallel patterns.

Many other projects also rely on the provision of implementation variants of functions, methods, or components for addressing performance and performance portability issues including PetaBricks [1], Merge [12], and Elastic computing [18].

PetaBricks addresses performance portability mostly across homogeneous multi-core architectures by focusing on auto-tuning methods for different types of optimization criteria. Merge also provides variants, but focuses on MapReduce as a unified, high-level programming model. Elastic computing provides so called elastic functions which represent a large number of variants, among which the best combination is composed guided by performance profiles and models.

PEPPHER also facilitates the use of auto-tuning techniques by exposing tunable parameters of both components and parameterized, adaptive library algorithms. Within the Autotune project [15], the PEPPHER framework has been coupled with the Periscope Tuning Framework, in order to tune pipeline patterns by dynamically selecting stage replication factors, buffer sizes (see Fig. 3), as well as to determine the best number of CPU cores and GPUs that should be used for a specific application run.

The EXCESS project builds on the PEPPHER component model and extends the approach with a focus on energy optimization by taking into account both software and hardware aspects [13].

7   Conclusion

In this paper we provided an overview of the PEPPHER methodology and framework for facilitating programmability and performance portability for single-node heterogeneous many-core architectures. Due to the increased complexity and performance variability exhibited by emerging parallel architectures, software for such systems needs to become more flexible in order to be portable across different systems and architecture generations. Achieving portability while ensuring programmability will require mechanisms that allow expressing the parallelism available in applications at a higher level of abstractions, while decoupling computations from their actual implementation and loci of execution. To support such approaches, intelligent runtime systems for dynamically managing parallel execution and adapting the granularity of parallelism to the actual hardware will become increasingly important.

References

1. Ansel, J., Chan, C.P., Wong, Y.L., Olszewski, M., Zhao, Q., Edelman, A., Amarasinghe, S.P.: PetaBricks: a language and compiler for algorithmic choice. In: Proceedings of the 2009 ACM SIGPLAN Conference on Programming Language Design and Implementation, PLDI 2009, pp. 38–49. ACM (2009)
2. Augonnet, C., Thibault, S., Namyst, R., Wacrenier, P.-A.: StarPU: a unified platform for task scheduling on heterogeneous multicore architectures. Concurrency Comput. Pract. Experience Spec. Issue: Euro-Par 23, 187–198 (2011)
3. Benkner, S., Bajrovic, E., Marth, E., Sandrieser, M., Namyst, R., Thibault, S.: High-level support for pipeline parallelism on many-core architectures. In: Kaklamanis, C., Papatheodorou, T., Spirakis, P.G. (eds.) Euro-Par 2012. LNCS, vol. 7484, pp. 614–625. Springer, Heidelberg (2012)

4. Benkner, S., Pllana, S., Träff, J.L., Tsigas, P., Dolinsky, U., Augonnet, C., Bachmayer, B., Kessler, C., Moloney, D., Osipov, V.: PEPPHER: efficient and productive usage of hybrid computing systems. IEEE Micro **31**(5), 28–41 (2011)
5. Bradski, D.G.R., Kaehler, A.: Learning OpenCV, 1st edn. O'Reilly Media Inc, Sebastopol (2008)
6. Bueno, J., Planas, J., Duran, A., Badia, R., Martorell, X., Ayguade, E., Labarta, J.: Productive programming of GPU clusters with OmpSs. In: Parallel Distributed Processing Symposium (IPDPS 2012), (2012)
7. Dastgeer, U., Li, L., Kessler, C.: The PEPPHER composition tool: performance-aware composition for GPU-based systems. Computing **96**(12), 1195–1211 (2014)
8. Hugo, A., Guermouche, A., Wacrenier, P.-A., Namyst, R.: Composing multiple StarPU applications over heterogeneous machines: a supervised approach. Int. J. High Perform. Comput. Appl. **28**, 285–300 (2014)
9. Kaiser, H., Heller, T., Adelstein-Lelbach, B., Serio, A., Fey, D.: HPX - a task based programming model in a global address space. In: PGAS 2014: The 8th International Conference on Partitioned Global Address Space Programming Models (2014)
10. Kessler, C., Dastgeer, U., Thibault, S., Namyst, R., Richards, A., Dolinsky, U., Benkner, S., Traff, J., Pllana, S.: Programmability and performance portability aspects of heterogeneous multi-/manycore systems. In: Design, Automation Test in Europe Conference Exhibition (DATE), pp. 1403–1408, March 2012
11. Lee, H.J., Brown, K., Sujeeth, A., Chafi, H., Olukotun, K., Rompf, T., Odersky, M.: Implementing domain-specific languages for heterogeneous parallel computing. IEEE Micro **31**(5), 42–53 (2011)
12. Linderman, M.D., Collins, J.D., Wang, H., Meng, T.H.Y.: Merge: a programming model for heterogeneous multi-core systems. In: Proceedings of the 13th International Conference on Architectural Support for Programming Languages and Operating Systems (ASPLOS 2008), pp. 287–296. ACM (2008)
13. Liu, L., Kessler, C.: Validating energy compositionality of GPU computations. In: Proceedings of the HiPEAC Workshop on Energy Efficiency with Heterogeneous Computing (EEHCO-2015) in conjunction with HiPEAC-2015 Conference, Amsterdam, The Netherlands (2015)
14. Mattson, T., Cledat, R., Budimlic, Z., Cave, V., Chatterjee, S., Seshasayee, B., van der Wijngaart, R., Sarkar, V.: OCR the Open Community Runtime Interface, version 1.0.0, June 2015
15. Miceli, R., Civario, G., Sikora, A., César, E., Gerndt, M., Haitof, H., Navarrete, C., Benkner, S., Sandrieser, M., Morin, L., Bodin, F.: Autotune: a plugin-driven approach to the automatic tuning of parallel applications. In: Manninen, P., Öster, P. (eds.) PARA. LNCS, vol. 7782, pp. 328–342. Springer, Heidelberg (2013)
16. Sandrieser, M., Benkner, S., Pllana, S.: Using explicit platform descriptions to support programming of heterogeneous many-core systems. Parallel Comput. **38**(1–2), 52–65 (2012)
17. Topcuoglu, H., Hariri, S., Wu, M.-Y.: Performance-effective and low-complexity task scheduling for heterogeneous computing. IEEE Trans. Parallel Distrib. Sys. **13**(3), 260–274 (2002)
18. Wernsing, J.R., Stitt, G.: Elastic computing: a framework for transparent, portable, and adaptive multi-core heterogeneous computing. In: Proceedings of the ACM SIGPLAN/SIGBED 2010 Conference on Languages, Compilers, and Tools for Embedded Systems (LCTES), pp. 115–124. ACM (2010)

Flexible Interpolation for Efficient Model Checking

Antti E.J. Hyvärinen[✉], Leonardo Alt, and Natasha Sharygina

Faculty of Informatics, Università della Svizzera italiana, Via Giuseppe Buffi 13,
CH-6904 Lugano, Switzerland
antti.hyvarinen@gmail.com

Abstract. Symbolic model checking is one of the most successful techniques for formal verification of software and hardware systems. Many model checking algorithms rely on over-approximating the reachable state space of the system. This task is critical since it not only greatly affects the efficiency of the verification but also whether the model-checking procedure terminates. This paper reports an implementation of an over-approximation tool based on first computing a propositional proof, then compressing the proof, and finally constructing the over-approximation using Craig interpolation. We give examples of how the system can be used in different domains and study the interaction between proof compression techniques and different interpolation algorithms based on a given proof. Our initial experimental results suggest that there is a non-trivial interaction between the Craig interpolation and the proof compression in the sense that certain interpolation algorithms profit much more from proof compression than others.

1 Introduction

Automated methods for formally verifying the absence of faults in a computer system are becoming increasingly important due to the significant role computers have in the society. Model checking [4] is one of the most successful approaches for formal verification. The underlying idea in model checking is to exhaustively explore a well-defined part of the state space of a system and either find errors or prove their absence in the studied state space. The problem is generally seen to be very hard and often undecidable, especially when the state space to be explored is the full state space of the system. To overcome the computational difficulty of verification many of the efficient approaches are based on describing the system using a logic-based formalism in which the lack of faults can be checked using efficient reasoning engines [3,6,7].

Many of the tools supporting traversal of the search space using logic-based, symbolic representation require methods for over-approximating parts of the state-space of the system being studied. A widely used approach is based on constructing Craig interpolants [5]. The idea is to partition an unsatisfiable logic formula into two parts $A \land B$ of which the A part needs to be over-approximated.

© Springer International Publishing Switzerland 2016
J. Kofroň and T. Vojnar (Eds.): MEMICS 2015, LNCS 9548, pp. 11–22, 2016.
DOI: 10.1007/978-3-319-29817-7_2

Craig interpolation provides a way of constructing an interpolant I which safely over-approximates A in the sense that $A \to I$ and $I \wedge B$ is still unsatisfiable.

This paper studies a framework for constructing propositional Craig interpolants through compressed resolution refutations and the labeled interpolation system [8]. The approach itself has been discussed in our previous work [1,12,13]; the novelty of this paper is in presenting the techniques under a uniform notation and reporting initial experimental results on combining the previously studied techniques.

The presented techniques have been implemented in the PeRIPLO interpolation engine http://verify.inf.usi.ch/periplo. The paper is organized as follows: Sect. 2 discusses approaches for symbolic model checking where interpolation has natural applications and introduces interpolation and our notation for propositional logic. Section 3 discusses the approach PeRIPLO uses for compressing the refutations it creates, and Sect. 4 discusses the PeRIPLO implementation of the labeled interpolation system. We report the experimental study in Sect. 5 and conclude in Sect. 6.

2   Preliminaries

Symbolic model checking consists of determining exhaustively whether the implementation of a system conforms to its specification. The system is defined as a finite set of variables $X = \{x_1, \ldots, x_n\}$ whose values change over discrete time $t = 0, 1, \ldots$ according to a transition relation T, and at time $t = 0$ satisfy the *initial condition* $I(X)$. The initial condition and the transition relation are defined as formulas over first order logic. An assignment $\sigma(X)$ mapping each variable in X to a concrete value is a *state* of the system. Given two copies of the system variables X and X' and two states $\sigma(X)$ and $\sigma'(X')$ the system can transition from $\sigma(X)$ to $\sigma'(X')$ from time t to $t+1$ if the assignments satisfy the transition relation $T(X, X')$. In this paper we consider specifications on the *safety* of a system: A system is safe if, whenever the system starts from a state satisfying the initial conditions and transitions according to the transition relation, the visited states $\sigma_0, \sigma_1, \ldots$ never satisfy the *error condition* $E(X)$ defined in the specification also as a formula in first order logic.

To show a system unsafe it sufficies to find a sequence of states $\sigma_0, \ldots \sigma_n$ satisfying

$$I(X_0) \wedge T(X_0, X_1) \wedge \ldots \wedge T(X_{n-1}, X_n) \wedge E(X_n). \tag{1}$$

To show a system safe one needs to find a formula $R(X)$ such that

$$\models I(X) \to R(X) \tag{2}$$

$$\models R(X) \wedge T(X, X') \to R(X'), \text{ and} \tag{3}$$

$$R(X) \wedge E(X) \text{ is unsatisfiable.} \tag{4}$$

The formula $R(X)$ above is the *safe inductive invariant* which is inductive by the second tautology and safe by the third formula. It is often more practical to interchange the roles of initial and error conditions since this will make the problem solving more incremental. In some algorithms this requires the definition of the inverse of the transition relation $T^{-1}(X, X')$.

In the following we will present two model-checking applications using this generic framework that will motivate our work on computing over-approximations: the *k-induction* for unbounded model checking, and *function summarization*. Finally we give the notation for propositional satisfiability and interpolation we will use in the paper.

k-Induction. A widely used algorithm for symbolic model checking is based on constructing the safe inductive invariant $R(X)$ by means of unrolling the transition relation k times, showing that the states reached after k steps do not satisfy the error condition, and trying to obtain a safe over-approximation of the initial condition based on the proof to heuristically compute an inductive invariant. This process is known as *k-induction*. To obtain the invariant in the form given in Eq. (2) the problem is stated as an over-approximation of the initial condition. In case of over-approximation of the final condition the resulting invariant will be safe in the sense that the inverse transition function T^{-1} cannot lead to a state satisfying the initial condition starting from a state satisfying the error condition. The critical part of this algorithm is the construction of the safe transitive invariant through over-approximation of the initial condition. A widely used approach for computing the over-approximation is through interpolation.

Function Summarization. In typical programming languages the programmer imposes a logical structure for a system by organizing the program into functions. From the perspective of model checking the functions offer an interesting approach for guiding the construction of the proof of correctness through *function summaries*.

Functions and their summaries are encoded into the transition function T modularly. Let program P have a function f, and let the encoding of the function f in logic be $|f|(X)$. If a proof of safety with respect to a verification condition c for a program is obtained, the function f can be over-approximated with respect to the verification condition in a safe way by replacing the encoding $|f|(X)$ with the over-approximating encoding $|\hat{f}_c|(X)$ that can still be used to prove correctness of the condition $c(X)$.

We mention two potential uses for this approach. The first is in verifying a sequence of verification conditions. Often the error condition $E(X)$ can be split in a natural way to several verification conditions $c_0, \ldots, c_n$ such that $E(X) = \vee_{i=0}^{n} c_i(X)$. Depending on the over-approximation and the relations between the conditions in the sequence it is often possible to organize the sequence so that the strong conditions are checked before the weak conditions. For instance [9] presents a heuristic for ordering verification conditions in a way that likely results in such a sequence. In this case an initial over-approximation can be used to verify the remaining conditions. The second application of function summaries is in verifying software upgrades. Given a function f and an upgraded function f', depending on the type of the upgrade it might be possible to avoid checking the compatibility of the new version of the software against the error condition. Instead of the expensive re-verification the check can be done locally by determining whether the safe over-approximation of the encoding $|\hat{f}_c|(X)$ contains the behavior of the upgraded function $|f'|(X)$, that is, by checking whether $|\hat{f}_c|(X) \rightarrow |F'|(X)$.

Interpolation and Propositional Satisfiability. All the above presented scenarios need an approach for constructing over-approximations of parts of the formula in Eq. (1). A widely used framework for this purpose is the Craig interpolation [5]. In this work we study in particular the *proof compression* and the *labeled interpolation system* [8] for propositional satisfiability.

Propositional satisfiability provides a convenient an expressive language for presenting instances of different model checking problems. Given finite set of Boolean variables B, the set of literals over B is $\{p, \neg p \mid p \in B\}$. A *clause* is a set of literals and a *formula in conjunctive normal form* (CNF) is a set of clauses. We use interchangeably the notation $\{l_1, \ldots, l_n\}$ and $l_1 \vee \ldots \vee l_n$, where l_i are literals, to denote clauses. Given a clause n, the set $vars(n) = \{p \mid p \in n \text{ or } \neg p \in n\}$ gives the variables of n.

A *resolution step* is a triple $n^+, n^-, (n^+ \cup n^-) \setminus \{p, \neg p\}$, where n^+ and n^- are two clauses such that $p \in n^+$, $\neg p \in n^-$, and for no other variable q both $q \in n^- \cup n^+$ and $\neg q \in n^- \cup n^+$. The clauses n^+ and n^- are called the *antecedents*, the latter is the *resolvent* and p is the *pivot* of the resolution step. A *resolution refutation* R of an unsatisfiable formula ϕ is a directed acyclic graph where the nodes are clauses and the edges are directed from the antecedents to the resolvents. The leaf nodes of a refutation R, i.e., nodes with no incoming edge, are the clauses of ϕ, and the rest of the clauses are resolvents derived with a resolution step. The unique node with no outgoing edges is the empty clause.

3   Methods of Proof Compression

A common approach for constructing interpolants is to compute a resolution refutation and label the refutation iteratively in a way that finally results in an interpolant. Since the resolution proof is often big and the interpolant size is one of the critical factors determining the usability of the interpolant it is preferable to obtain as small interpolants as possible. This section gives an in-depth view of the techniques implemented in the PeRIPLO tool for compressing the refutation once it has been constructed. In particular we cover the *local transformation framework*, the *pivot recycling algorithm* and an approach for delaying resolution steps that involve a unit clause as an antecedent.

The local transformation framework. In our experiments an important factor making resolution proofs of SAT solvers big is that the solver often resolves on a pivot several times. This type of redundancy can always be removed from a refutation and the resulting refutation will remain sound. The local transformation framework [13] addresses this issue. The framework consists of two rules for reducing a proof, complemented with two reshuffling rules that are employed to

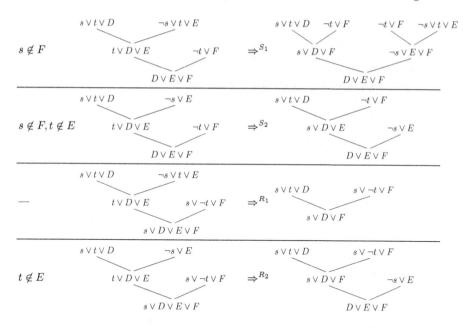

Fig. 1. Transformation rules

give more opportunities for the application of the two reducing rules. The four reduction and reshuffling rules are presented in Fig. 1. The restrictions on the application on the rule are listed on the leftmost column.

The PeRIPLO system uses the local transformation system to detect and remove redundancies from a refutation. The algorithm for applying the system is given in Fig. 2. The critical part of the algorithm is on lines 8–9 where the algorithm identifies a *context*, an environment which matches one of the rules in Fig. 1. The context consists of the two pivots p and q and the surrounding clauses, and since the system is symmetric the resolvents have both *left* and *right* contexts. Once a context is found the algorithm applies heuristically one of the transformation rules on the context on line 10.

The use of the transformation rules might render the refutation invalid if a clause is a resolvent in more than one resolution steps. To avoid the problem the rules R_1 and R_2 are not used in such cases. Finally the lines 12, 14, and 16 take care of the cases where resolution step has become useless due to the compression.

The pivot recycling algorithm. While the removal of the doubly appearing pivots can be done with the proof transformation system of Fig. 1, it is often useful to combine the approach with a more aggressive approach based on reachability on the refutation. One way of implementing the safe removal of extra resolution steps in a refutation DAG is to prevent the removal operation on resolvents that are used in more than one resolution step. However this approach is too

Input : R — A refutation

      T — A time limit

Output : R' — a compressed refutation

1 **while** T is not surpassed:

2    $TS :=$ topologically sorted list of clauses in R

3   **for** $n \in TS$:

4   **if** n is not a leaf:

5            $p := piv(n)$

6   **if** $\neg p \in n^-$ and $p \in n^+$:

7                $n := (n^- \cup n^+) \setminus \{p, \neg p\}$

8                $lc :=$ left context of n

9                $rc :=$ right context of n

10               $ApplyRule(lc, rc)$

11   **else if** $p \in n^+$:

12   Substitute n with n^-

13   **else if** $\neg p \in n^-$:

14   Substitute n with n^+

15   **else**

16   Heuristically choose either n^+ or n^- and substitute n with it

Fig. 2. The local transformation framework algorithm.

restrictive since often the literals are resolved on other paths as well. For this purpose PeRIPLO uses the recycle pivots with intersection algorithm, presented in [10] and based on the original recycle pivots algorithm of [2]. We present an implementation adapted from [13] in Fig. 3, designed for a slightly more general case where the root of the refutation might contain a non-empty clause. The algorithm takes as input a refutation R and computes for each clause n in R the set of literals that can be safely removed from the literal into the set RL. The respective literals in $n \cap RL[n]$ are then removed from n and the algorithm guarantees that the refutation can be transformed to a valid refutation afterwards. The critical reasoning is done on lines 14, 20, 25, and 29 where the information on which literals can be removed on other paths where a resolvent n is resolved is used to refine the removable literals for its parents n^- and n^+.

Delaying unit resolution. A good heuristic for reducing the size of the refutation is to move the resolution steps where one of the resolvents is a unit clause to the root. This is useful since it gives a natural way of guaranteeing that the unit clauses are resolved only once in the refutation. The PeRIPLO solver implements this idea as the *PushdownUnits* algorithm [13] by identifying sub-refutations that end in a unit clause, detaching them from the refutation and, if necessary, attaching them above the resolution step resulting in the root.

The PeRIPLO proof compression algorithm. The PeRIPLO system uses an approach for proof compression that combines both the pivot recycling algorithm presented in Fig. 3 and the proof reduction framework (Fig. 2). The hybrid algorithm is presented in Fig. 4. The algorithm calls as the first step the procedure

Input : R — A refutation
Output : RL — the mapping from resolvents to the literals that can be removed in them
1   $TS :=$ topologically sorted list of clauses in R
2   $RL := \emptyset$ // The set of removable literals
3 **for** $n \in TS$:
4   **if** n is not a leaf:
5   **if** n is the root:
6           $RL[n] := \{\neg p \mid p \in n\}$
7   **else**:
8           $p := piv(n)$
9   **if** $p \in RL[n]$:
10             $n^+ := null$
11   **if** n^- not seen yet:
12                $RL[n^-] := RL[n]$
13   Mark n^- as seen
14   **else** $RL[n^-] := RL[n^-] \cap RL[n]$
15   **else if** $\neg p \in RL[n]$:
16             $n^- := null$
17   **if** n^+ not seen yet:
18                $RL[n^+] := RL[n]$
19   Mark n^+ as seen
20   **else** $RL[n^+] := RL[n^+] \cap RL[n]$
21   **else if** $p \notin RL[n]$ and $\neg p \notin RL[n]$:
22   **if** n^- not seen yet:
23                $RL[n^-] := RL \cup \{p\}$
24   Mark n^- as seen
25   **else** $RL[n^-] := RL[n^-] \cap (RL[n] \cup \{p\})$
26   **if** n^+ not seen yet:
27                $RL[n^+] := RL[n] \cup \{\neg p\}$
28   Mark n^+ as seen
29   **else** $RL[n^+] := RL[n^+] \cap (RL[n] \cup \{\neg p\})$
30 **return** RL.

Fig. 3. The *RecyclePivotsWithIntersection* algorithm

Input : R — A refutation;
　　　　I — the number of loop iterations;
　　　　T — A time limit for the proof reduction framework
Output : R' — A compressed refutation
1   $R' := PushdownUnits(R)$
2 **for** $i = 0$ to I
3     $R' := RecyclePivotsWithIntersection(R')$
4     $R' := ReduceAndExpose(R', T)$
5 **return** R'.

Fig. 4. The hybrid algorithm for proof compression

for moving unit resolutions to the root of the refutation and then repeatedly calls the functions *RecyclePivotsWithIntersection* and *ReduceAndExpose* to gradually obtain a more compact proofs.

4   Labeling in Interpolation

The PeRIPLO interpolation algorithm is based on computing the propositional interpolant from a refutation and a labeling which allows tuning of the interpolant to specific needs and the refutation. The implementation is based on the *labeled interpolation system* originally presented in [8] (LIS) and further developed in [1].

The system works on an *interpolation instance* (R, A, B), where R is the refutation of $A \wedge B$ and A is the formula to be over-approximated. Given a clause n in R and a variable $p \in vars(n)$ occurring in the clause, the system assigns a unique label $L(p, n)$ from the set $\{a, b, ab\}$ to the occurrence (p, n). In the leaf clauses the labeling is restricted to $L(p, n) = a$ if $p \notin vars(B)$ and $L(p, n) = b$ if $p \notin vars(A)$, but can be freely chosen for leaf occurrences of variables in $vars(A) \cap vars(B)$. In the resolvent clauses n_r of R the labeling $L(p, n_r)$ is determined by the label in n^+ and n^-. If $p \in vars(n^+) \cap vars(n^-)$ and $L(p, n^+) \neq l(p, n^-)$, then $L(p, n) = ab$, and in all other cases the label of the occurrence $L(p, n)$ is either $L(p, n^+)$ or $L(p, n^-)$.

The final interpolant is constructed based on the labeling and the refutation R iteratively for each clause in R starting from the leaf clauses and ending in the root. In particular, for a leaf clause n_l the interpolant is

$$I(n_l) = \begin{cases} \bigvee\{p \mid p \in n_l \text{ and } L(vars(p), n_l) = b\} & \text{if } n_l \in A, \text{ and} \\ \bigwedge\{\neg p \mid p \in n_l \text{ and } L(vars(p), n_l) = a\} & \text{if } n_l \in B \end{cases} \quad (5)$$

The partial interpolant of a resolvent clause n_r with pivot p and antecedents n^+ and n^-, where $p \in n^+$ and $\neg p \in n^-$, is

$$I(n_r) = \begin{cases} I(n^+) \vee I(n^-) & \text{if } L(p, n^+) = L(p, n^-) = a, \\ I(n^+) \wedge I(n^-) & \text{if } L(p, n^+) = L(p, n^-) = b, \text{ and} \\ (I(n^+) \vee p) \wedge (I(n^-) \vee \neg p) & \text{otherwise.} \end{cases} \quad (6)$$

Several different approaches for constructing efficient labelings have been proposed. These include approaches for logically strong and weak interpolants [8,11], and our recent work on proof-sensitive labelings [1].

5   Experiments

We report here experimental results on both the proof compression approaches discussed in Sect. 3 and the labeled interpolation system of Sect. 4 using the PeRIPLO tool. Figure 5 shows the architecture of the tool. The experiments use a very basic form of proof compression where the algorithm in Fig. 4 uses the iteration count $I = 1$ and does not run *ReduceAndExpose* on line 4. The interpolator is used with six different interpolation algorithms: the weakest and the strongest interpolants M_w, M_s available from the LIS; three versions PS_w, PS, PS_s of a labeling function that attempt to minimize the interpolant size by labeling occurrences so that the minimum number of literals appear in the partial interpolants

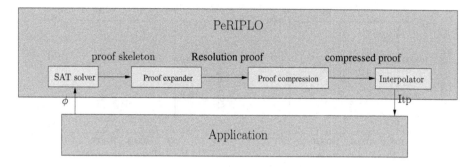

Fig. 5. The PeRIPLO architecture.

Table 1. Relative compression efficiency for different labeling functions.

	M_w	PS_w	P	PS	PS_s	M_s
time	3.25	3.10	2.77	2.19	2.10	2.20
size	15.15	15.86	12.04	7.20	6.57	6.13

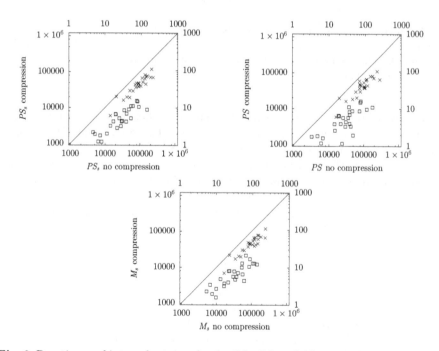

Fig. 6. Run time and interpolant sizes for the PS_s, PS, and M_s interpolation algorithms with and without proof compression. The left and bottom axes are in bytes and the top and right axes are in seconds. The $\times$ and $\square$ symbols are, respectively, the time and the size measurements.

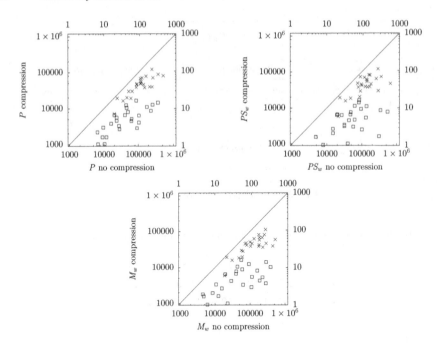

Fig. 7. Run time and interpolant sizes for the P, PS_w, and M_w interpolation algorithms with and without proof compression. The left and bottom axes are in bytes and the top and right axes are in seconds. The $\times$ and $\square$ symbols are, respectively, the time and the size measurements.

of the leaves; and P, an algorithm that labels all occurrences with ab. The experiments are done using the FunFrog system as the application. The system employs function summarization as explained in Sect. 2.

Table 1 reports the relation of size and time between the interpolants resulting from the algorithm without and with proof compression over roughly 25 interpolation instances. In general the proof compression helps in reducing both the size of the interpolant and the time required to construct the interpolant. The reduction in size is more significant than in run time. This is not unexpected since the run time contains several constant elements such as the time required to solve the instance. Interestingly the efficiency of the proof compression is not the same for all the interpolation algorithms. The algorithms M_w, PS_w, and P profit significantly more from the compressed refutation than the other algorithms. These algorithms often produce bigger interpolants, and there seems to be a non-trivial interaction between how the proof is reduced and how the different labeling functions are able to use the smaller proof.

We report the individual results as scatter plots in Figs. 6 and 7 for both time ($\times$) and size ($\square$). The results show a consistent reduction in all cases but also show several cases where the compression results in two orders of magnitude reduction in size.

6   Conclusions

This paper presents a range of applications in model checking where interpolation plays a critical role. We present two major techniques that affect the efficiency of interpolation: proof compression and labeling. Both techniques are described in detail showing how they are implemented in the propositional interpolation tool PeRIPLO. Finally we analyze the effect of proof compression when combined with different interpolant labellings on one of the applications. We reveal an interesting behavior that not all labeling functions profit in the same way from the proof compression. This suggests a non-trivial interaction between the interpolation and the proof compression that requires further studying.

Currently we are planning to extend the ideas presented in this paper to Satisfiability Modulo Theories in general, and applying them in other novel application domains where interpolation is useful.

Acknowledgements. This work has been supported by the SNF project numbers 200020_163001 and 200021_153402, and the ICT COST Action IC1402.

References

1. Alt, L., Fedyukovich, G., Hyvärinen, A.E.J., Sharygina, N.: A proof-sensitive approach for small propositional interpolants. In: Proceedings of VSTTE 2015 (2015, to appear)
2. Bar-Ilan, O., Fuhrmann, O., Hoory, S., Shacham, O., Strichman, O.: Linear-time reductions of resolution proofs. In: Chockler, H., Hu, A.J. (eds.) HVC 2008. LNCS, vol. 5394, pp. 114–128. Springer, Heidelberg (2009)
3. Biere, A., Cimatti, A., Clarke, E., Zhu, Y.: Symbolic model checking without BDDs. In: Cleaveland, W.R. (ed.) TACAS 1999. LNCS, vol. 1579, p. 193. Springer, Heidelberg (1999)
4. Clarke, E., Emerson, E.: Design and synthesis of synchronization skeletons using branching-time temporal logic. In: Kozen, D. (ed.) Logic of Programs 1981. LNCS, vol. 131, pp. 52–71. Springer, Heidelberg (1982)
5. Craig, W.: Three uses of the Herbrand-Gentzen theorem in relating model theory and proof theory. J. Symbolic Logic **22**(3), 269–285 (1957)
6. Davis, M., Putnam, H.: A computing procedure for quantification theory. J. ACM **7**(3), 201–215 (1960)
7. Detlefs, D., Nelson, G., Saxe, J.B.: Simplify: a theorem prover for program checking. J. ACM **52**(3), 365–473 (2005)
8. D'Silva, V., Kroening, D., Purandare, M., Weissenbacher, G.: Interpolant strength. In: Barthe, G., Hermenegildo, M. (eds.) VMCAI 2010. LNCS, vol. 5944, pp. 129–145. Springer, Heidelberg (2010)
9. Fedyukovich, G., D'Iddio, A.C., Hyvärinen, A.E.J., Sharygina, N.: Symbolic detection of assertion dependencies for bounded model checking. In: Egyed, A., Schaefer, I. (eds.) FASE 2015. LNCS, vol. 9033, pp. 186–201. Springer, Heidelberg (2015)
10. Fontaine, P., Merz, S., Woltzenlogel Paleo, B.: Compression of propositional resolution proofs via partial regularization. In: Bjørner, N., Sofronie-Stokkermans, V. (eds.) CADE 2011. LNCS, vol. 6803, pp. 237–251. Springer, Heidelberg (2011)

11. McMillan, K.L.: An interpolating theorem prover. In: Jensen, K., Podelski, A. (eds.) TACAS 2004. LNCS, vol. 2988, pp. 16–30. Springer, Heidelberg (2004)
12. Rollini, S.F., Alt, L., Fedyukovich, G., Hyvärinen, A.E.J., Sharygina, N.: PeRIPLO: a framework for producing effective interpolants in SAT-based software verification. In: McMillan, K., Middeldorp, A., Voronkov, A. (eds.) LPAR-19 2013. LNCS, vol. 8312, pp. 683–693. Springer, Heidelberg (2013)
13. Rollini, S.F., Bruttomesso, R., Sharygina, N., Tsitovich, A.: Resolution proof transformation for compression and interpolation. Formal Methods Syst. Des. **45**(1), 1–41 (2014)

Understanding Transparent and Complicated Users as Instances of Preference Learning for Recommender Systems

P. Vojtas[(⊠)], M. Kopecky, and M. Vomlelova

Faculty of Mathematics and Physics, Charles University in Prague,
Malostranske Namesti 25, Prague, Czech Republic
{vojtas,kopecky}@ksi.mff.cuni.cz,
marta@ktiml.mff.cuni.cz

Abstract. In this paper we are concerned with user understanding in content based recommendation. We assume having explicit ratings with time-stamps from each user. We integrate three different movie data sets, trying to avoid features specific for single data and try to be more generic. We use several metrics which were not used so far in the recommender systems domain. Besides classical rating approximation with RMSE and ratio of order agreement we study new metrics for predicting Next-k and (at least) 1-hit at Next-k. Using these Next-k and 1-hit we try to model display of our recommendation – we can display k objects and hope to achieve at least one hit.

We trace performance of our methods and metrics also as a distribution along each single user. We define transparent and complicated users with respect to number of methods which achieved at least one hit.

We provide results of experiments with several combinations of methods, data sets and metrics along these three axes.

Keywords: Content based recommendation · Explicit ratings · Integration of three movie datasets · User understanding · User preference learning · RMSE · Next-k · 1-hit · Order agreement metrics

1 Introduction

In this paper we are concerned with user understanding in content based recommendation displayed on the k-objects sized screen. We assume having a user-item matrix with explicit ratings and time-stamps. Our tasks vary from estimating subjective ordering of items based on either explicit or synthetic attributes of items to learning/predicting user preferences on unseen/unrated items.

In this paper we are not going to present any new mining techniques nor improve achievements of comparable efforts on same data and metrics. Rather, we concentrate on three issues:

First, we integrate three different data sets from the same domain of movie recommendation, trying to avoid features specific for single data and try to be more generic. We consider also semantic enrichment of movie data to enable content based recommendation.

© Springer International Publishing Switzerland 2016
J. Kofroň and T. Vojnar (Eds.): MEMICS 2015, LNCS 9548, pp. 23–34, 2016.
DOI: 10.1007/978-3-319-29817-7_3

Second, we use several metrics which were not used so far in the recommender systems domain. Besides classical rating approximation with RMSE and top-k we study new metrics for predicting Next-k and (at least) 1-hit.

Finally and most importantly, we trace performance of our methods and metrics as a distribution along each single user. This helps us to detect transparent and complicated users. Improving these metrics (for instance 1-hit) can contribute to the increase of user satisfaction (we display Next-k and hope to achieve at least 1-hit).

We provide results of experiments with several combinations of methods, data sets and metrics along of these three axes. Nevertheless, all our experiments are offline on public accessible data. A validation in real online A/B testing needs access to an application. We were not able to realize this so far. We concentrate only on algorithmic part of recommendation, business and marketing part of recommendation problem is out of scope of this study.

In the area of recommender systems same-data-challenges probably the most famous is the *Netflix challenge*. Results were measured by improvement in RMSE of ranking prediction. In [7] we have described our participation in ESWC 2014 *Linked Open Data-enabled Recommender Systems Challenge*. In [8, see also 9] we described our approach to *Movie Tweets Popularity Challenge* at ACM-RecSys 2014. In [11] we presented some initial results on enriched *MovieLens* data sets in the *RuleML challenge* competition. In [6] we presented a more detailed analysis of results in [11].

We try to base our recommendation on rules to provide human understandable explanation. For rules we use GAP – generalized annotated logic programming rules of [5]. Induction of many valued rule systems was handled in [4]. Nevertheless data sets for induction were small and we did not consider recommendation for multiple users at that time.

Prediction of recommendation/ranking for multiple users started to be popular also with the *Yahoo Challenge*. For our purpose we found relevant mining tricks from [1] and winners of 2014 ACM *RecSys Challenge* [3]. One of main motivation for us was presentation Kate Smith-Miles at EURO 2014 ([10]), where difficulty of instances of graph colouring was modelled according to number of algorithms which found the instance difficult. Here, instead of graphs, we consider difficulty of users.

Last but not least we are following the CRISP-DM – the *Cross Industry Standard Process for Data Mining* methodology described in [2]. CRISP-DM consists of following steps: *Business Understanding, Data Understanding, Data Preparation, Modelling, Evaluation* and *Deployment* (this will not be considered here).

1.1 Main Contribution of This Paper Are

Main contributions of this paper are:

- Data integration of three different movie data sets
- Distinguishing between preferences represented as function and as ordering – this influences both learning and metrics (function approximation/order agreement).
- All our experiments are offline – nevertheless to make our recommendation more realistic we consider new metrics *Next-k2 l* (recommending k elements we try to hit

l element test set, which is – according to time stamps – "next") and 1-hit (*Next-k2* l). We discuss our results also as distributed along users– this gives more insight than global averages of user's achievements

- We define transparency/difficulty of a user by number of methods where at least 1-hit into target set was achieved.

Paper is organized according to CRISP-DM categories: *Business Understanding, Data Understanding, Data Preparation, Modelling* and *Evaluation.*

2   Data Understanding, Data Preparation, Cleaning, Sampling, Integration

We analysed samples of three datasets from movie domain, referred as *MovieLens* (extracted from *MovieLens1 M* data), *Flix* (a sample of the famous competition enriched by movie titles), and *MovieTweets* (extracted from *RecSysChallenge2014* – from tweets table we used only time stamps). We restricted all three datasets to movies for that we have IMDB identifier to be able to join the datasets and enable content based recommendation. Further we restricted all three datasets to those movies we have *DBPedia* attributes available from *RuleML Challenge* [12].

After these restrictions data sets substantially differ in the number of rated movies for a user. Most ratings per user we have in the *Flix* dataset. On the other hand, in *MovieTweets* dataset only 42 users (out of 1547) rated more than 10 movies. To keep similar evaluation criteria for *Next-k* and 1-hit (explained later), we restricted the *MovieTweets* dataset to these 42 users for those two experiments. Distribution of users' ratings is depicted in Fig. 1.

MovieLens	*Flix*	*MovieTweets*
1000 users	499 users	170 out of 1547 users rated 5 or more movies

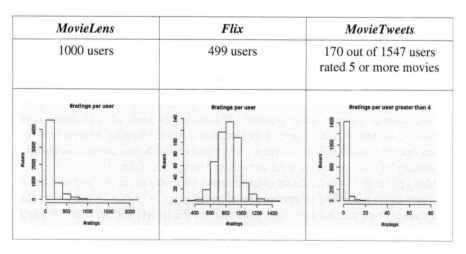

Fig. 1. Number of ratings distribution in our testing datasets – *MovieLens, Flix* and *MovieTweets*

Furthermore we use the movie attributes obtained from the *DBpedia* ([12]). These data were available in form of 5 003 binary flags like *Films_set_in_New_York_City*, *Screenplays_by_James_Cameron*, *Films_shot_in_Arizona*, *Film_scores_by__John_Williams*, etc. Our first guess when we started analysing *MovieLens* dataset was to take most frequent properties and check their influence of our prediction the most used flags relevant to this dataset are shown in Table 1.

Table 1. Some of most frequent properties (out of 5 003 *DBPedia* properties)

Order number	Category name	Number of movies
1	*English-language_films*	2104
2	*American_films*	1188
3	*Directorial_debut_films*	371
4	*1990s_drama_films*	344
...	*Warner_Bros._films*	294
8	*Films_set_in_New_York_City*	278
51	*Miramax_Films_films*	103

We saw that these do not influence our achievements on the task significantly. The relevance of most of these attributes on our task was very small. We can also see that most of those flags do not seem to influence user's preference.

Our second attempt was based on observation that certain types of properties repeat. In Table 2 we see most frequent repetitions in properties wording. Nevertheless, this did not bring us much further either.

Table 2. Many properties repeat, form clusters, nevertheless do not influence the task.

Repeated part of property	Number of properties
Films_directed_by_...	995
1757-2032...	443
Films_set_in_...	364

Third attempt brought some progress. We started to look what is inside those attributes – the rest of URI's were considered as natural language expressions. We counted occurrences of individual values of properties in all movies and compared it with number of occurrences of these movies in ratings (see Table 3).

Other approach was to create set of explanatory attributes of movies and set them to 1, respectively 0 according to appearance of some word or phrase in the movie flag descriptions. So we created explanatory attribute SPIELBERG and set it to 1 for each movie directed, produced or any other way connected to Steven Spielberg, similarly we created attribute LA and assigned it to all movies located to LA etc.

Table 3. *RatingCNT* is the number of movies in our internal candidate set with respective property. *MovieCNT* is a number of movies with this property. R/M is 1000-times ratio of these two numbers and last column is the Excel formula for weighted average between *MovieCNT* and R/M ratio (we show only an initial segment).

A PROP	B VALUE	C Rating CNT	D Movie CNT	E R/M	IF(D>3; 3*E+C; 0)
Films_set_in	*New_York_City*	18259	278	657	20230
Films_shot_in	*California*	13193	149	885	15848
Films_set_in	*Los_Angeles,_Calif**	12460	160	779	14797
Films_shot_in	*Los_Angeles,_Calif**	10025	121	829	12512
Films_that_won	*best_Sound_Edit*_Aw**	5369	24	2237	12080
Films_shot_in	*New_York_City*	9239	128	722	11405
Films_based_on	*novels*	9548	232	412	10784

3 Modelling, Methods, Feature Engineering

3.1 Evaluation Metrics

We studied selected models against different evaluation criteria. The basic task was to predict movies the user is going to watch next. We "hide" 10 last records according to the timestamp as test data, and the trained our model on remaining data.

Next-k from l – size and precision. Based on the model, we recommended $k=5$ and $k=10$ movies and tested the recommendation against $l=10$ hidden records. The Table 6 below presents the number of correct recommendations. From this, the precision can be calculated.

1-hit. We tried to recommend at least one correct movie for any user. In this metric, it does not matter whether the user got one or five good predictions – either he got some, or not. Again, percentage of met user can be calculated.

RMSE, mean absolute error (MAE). We trained linear regression model for each dataset to predict Rating. These models can be used for ordering movies and recommending the ones with the highest rating.

Ordering (dis)agreements. Completely different approach was not to approximate user ratings by some function, but understand them as a partial ordering of movies with respect to given user. If the user rated movie A lower than the movie B, we understood this fact as user's verdict that A is less than B. If the user rated both movies A and B the same, we have taken it as information, that A is equal to B. Then we tried to create different types of movie orderings based mainly on explanatory attributes, and observe the rate of agreement, respectively disagreement between such estimated ordering and the user's opinion.

3.2 Algorithms

Linear regression models. We trained linear regression models to predict Rating. Concerning movie attributes we only started to deal seriously with them. In our first experiments at the beginning they appeared not to be very important and so we have chosen simple models based on following input attributes:

The most relevant input attributes are following:

- **MovieAvg,** an average movie rating over training data. This was the most important attribute for *MovieLens* and *MovieTweets* datasets, the second most important for *Flix* data.
- **UserShift,** an average difference for ratings of given user related to average movie ratings. This attribute was the most important for *Flix* data (the data with the most ratings per user), it was also important in *MovieTweets* dataset, while mostly unimportant in the *MovieLens* dataset.
- **GenreMatch,** for any user, we have found the most watched genre. If a movie genre corresponds to user genre, this attribute is set to 1 and to 0 otherwise. This attribute was important in *MovieLens* and *Flix* datasets, while it was not important in *MovieTweets* dataset (where the user genre was set from a very small set of training examples, usually 1-3).

The attribute coefficients for different datasets are listed in the Table 4.

Table 4. Linear regression model trained on same attributes (GenreMatch is present for all data sets thanks to data integration)

Linear Regression Model coefficients				
	MovieAVG	userAVGshift	GenreMatch	Intercept
MovieTweets	0.900	0.293	0.000	0.808
MovieLens	0.997	0.000	0.062	−0.019
Flix	1.003	1.002	0.076	−0.035

Models for Next-k prediction. For Next-k prediction, we ordered movies according some criteria ("models") and recommended the top k movies. We used different models to order movies:

- As a basis, ordering according MovieAVG rating (the same for all users) was used. In some datasets, we have only small number of ratings for a movie, therefore we used also Bayesian estimate that makes the estimate more stable by prior probability for the overall estimate AVG, precisely: bayesAvg=(AVG*50+MovieAVG*MovieCNT)/(MovieCNT+50). Third choice was to use the linear regression model for each specific dataset.
- We tried specific algorithms for prediction described in ([1]). GENREMATCH prefers movies that belong to the genre mostly rated by the specific user. Matched movies are ordered according to bayesAVG. SPIELBERG and KSI models use specific movie attributes, Spielberg and Original the first one, KSI uses attributes listed in Table 6.

4 Experiments, Evaluation

4.1 Evaluation of Algorithms

We present Tables 5, 6 and 7 summarizing the evaluation measures for different datasets and methods. It shows kind of similarity of *MovieLens* and *MovieTweets* datasets. The algorithm SPIELBERG (recommend a movie if Spielberg is mentioned) gives good results. Instead of taking avgRating we added 50 hypothetical users voting for the overall average ([1]). For large data samples, like in *Flix* datasets, this Bayesian estimate bayesAVG does not change the estimate much. For small data samples, like *MovieLens* and *MovieTweets*, this method makes the estimate more stable.

Table 5. Results for *MovieLens* dataset

Method		Maximum	KSI-10 att.	Spielberg	Genre Match +bayes AVG	avg Rating	bayes AVG	Linear Regression ML
k=5	Next-k-size	5000	378	**559**	309	7	*195*	6
	Next-k-P		7.56%	**11.18%**	6.18%	0.14%	*3.90%*	0.12%
	1-hit-size	1000	371	**487**	256	7	*172*	6
	1-hitUserRatio		37.10%	**48.70%**	25.6%		*17.20%*	
k=10	Next-k-size	10000	415	**728**	585	7	*351*	8
	Next-k-P		4.15%	**7.28%**	5.85%		*3.51%*	0.08%
	1-hit-size	1000	394	**582**	389	7	*291*	

Table 6. Results for *Flix* dataset

Method		Maximum	KSI-10 att.	G.M.+bayes AVG	avg Rating	bayes AVG
k=5	Next-k-size	2495	51	160	*232*	232
	Next-k-P		2.04%	6.41%	*9.30%*	9.30%
	1-hit-size	499	50	134	*197*	192
k=10	Next-k-size	4990	96	4.81%	*7.23%*	6.91%
	Next-k-P		1.92%	183	*258*	243
	1-hit-size	499	90	160	*32*	232

Table 7. Results for *MovieTweets* dataset

Method		Maximum	KSI-10 att.	Spielberg	G.M. +bayes AVG	avg Rating	bayes AVG
k=5	Next-k-size	210	8	**10**	1	2	*9*
	Next-k-P		3.81%	**4.76%**	0.48%	0.95%	*4.29%*
	1-hit-size	42	7	**7**	1	2	*8*
k=10	Next-k-size	420	13	14	10	4	*19*
	Next-k-P		3.10%	**3.33%**	2.38%	0.95%	*4.52%*
	1-hit-size	42	10	**10**	8	4	*12*

Table 8. Most important explanatory attributes in *MovieLens* calculated as order of agreement. Due to the data integration these attributes can be studied in other data sets

Explanatory attribute	ML Sharp Agree	ML Agree	Flix Sharp Agree	Flix Agree	Movie Tweets Sharp Agree	Movie Tweets Agree
SOUNDMIX	**0.495**	**0.781**	0.527	0.819	0.527	0.807
SOUNDEDIT	0.489	0.779	0.544	**0.836**	**0.537**	**0.827**
SPIELBERG	0.491	0.772	**0.548**	0.829	0.426	0.752
ORIGINAL	0.475	0.766	0.469	0.778	0.461	0.810
VISUAL	0.461	0.750	0.502	0.801	0.481	0.773
CAMERON	0.435	0.728	0.451	0.764	0.510	0.706
WILLIAMS	0.430	0.716	0.465	0.766	0.402	0.711
MAKEUP	0.399	0.696	0.460	0.765	0.510	0.822
NOVELS	0.381	0.688	0.355	0.682	0.338	0.692
CALIF	0.377	0.670	0.363	0.682	0.363	0.657
NY	0.359	0.655	*0.345*	*0.670*	0.352	0.629
LA	0.359	0.650	0.350	*0.670*	0.369	0.638
ARIZONA	*0.348*	*0.639*	0.356	0.677	*0.286*	*0.563*

We predicted k records and compare the agreement with $l=10$ hidden last records (per user). For *Flix* dataset, and only for it, the linear regression method trained to predict rating gives good results also for prediction next movies.

4.2   Comparison of Movie Ordering Estimations

First, we taken thirteen different explanatory attributes, extracted from the DBPedia: ARIZONA, CALIF, CAMERON, LA, MAKEUP, NOVELS, NY, ORIGINAL, SOUNDEDIT, SOUNDMIX, SPIELBERG, VISUAL, WILLIAMS and for each of them we defined the movie ordering according to value of given explanatory attribute. If the movie A had given explanatory attribute set to 0, while the movie B had set it to 1, we define it as A is less than B etc. Then we took all combinations of $[u, A, B]$ where user u rated both movies and our estimate was $A<B$. If the A was less than B also according to user's ratings, we took it as a sharp agreement with our estimate. If the user's rating were equal, we still count it together with the previous case as an (not sharp) agreement. Then we were interested in the ratio of sharp agreements and agreements to the count of all combinations. Results, ordered in descending order according to agreement ratio in the *MovieLens* dataset are present in Table 6. The best results are highlighted by bold font, while the worst ones are in italic.

Further we wondered, if the combination of more explanatory attributes could improve our ordering estimate, and so we tried to define another orderings based not on individual attributes, but on vectors of first k of attributes from the previous table. So we assigned the k-dimensional vector of zeros and ones to each movie, and declare movie A less than movie B if and only if each A's vector component was less or equal than corresponding B's vector component, and at least one dimension was sharply ordered. The results didn't show improvement over usage of individual attributes, as it is shown on 10-dimensional vector results in Table 9.

Table 9. Agreement for 4-dimensional and 10-dimensional vectors of explanatory attributes

Explanatory attribute vector	ML Sharp Agree	ML Agree	Flix Sharp Agree	Flix Agree	Movie Tweets Sharp Agree	Movie Tweets Agree
<SMIX,..., ORIG>	0.397	0.691	0.381	0.704	0.405	0.697
<SMIX,..., CALIF>	0.377	0.670	0.363	0.682	0.363	0.657

4.3 Evaluations of Users

We analysed the distribution of evaluation criteria for different users. The users are displayed as circles or aggregated into a boxplot. The Fig. 2 shows a boxplot of *Agree* measure on y-axis for different algorithms on x-axis. We depicted two single attribute models (see top rows Table 8) and two multiple attribute models giving the highest mean in at least one of datasets. Their results are displayed for all three datasets.

We compare the prediction success of models with RMSE and *Sharp agree* criteria. On x-axis, a transparency of users for prediction models is displayed. The most transparent user (right) got at least one prediction from any model, the difficult users (left) did not get any correct prediction from any model. The RMSE error on y-axis shows only very weak dependence on the prediction transparency. The Sharp agree ratio (right part in Fig. 3) slightly increases with increasing transparency.

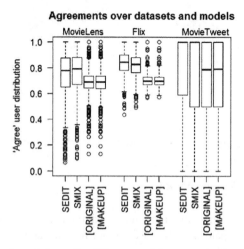

Fig. 2. User distribution of *Agree* for different models on x-axis, datasets above the graph.

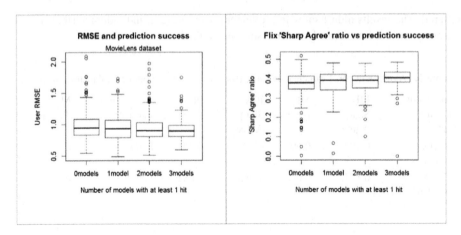

Fig. 3. Left: Distribution of RMSE (left) and *Sharp Agree* (right) between users distinguished by number of methods by which the user has at least one hit.

Figure 4 visualizes a complex model from Table 6. Circles correspond to users, *x* axis to *Sharp Agree* of ordering (left) and *Agree* (right), *y* axis shows RMSE, colours correspond to number of models with at least 1 good prediction. Triangles denote users for whom all 4 models give at least 1 good prediction. RMSE- is error measure (-), Agree is positive (+), the best position is the lower right corner of the graph.

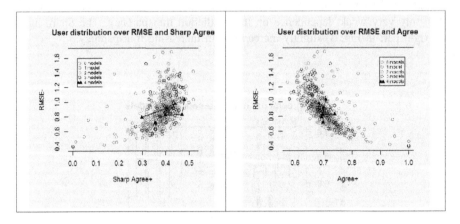

Fig. 4. Comparison of three success/error measures on *Flix* data

5   Conclusions, Future Work

For data integration of three different movie data sets and restriction to DBPedia attributes from [12] we paid the price: we lost lot of data samples.

We presented two views to preferences: these can be understood as either function or as ordering – this influences both learning and metrics – results are interesting and need a deeper insight. Nevertheless it is a challenge for future work.

One of most interesting part we find our experiments on Next-k (1-hit(Next-k)). This makes our recommendations more realistic as it is directly connected to what we display to the user. Moreover it shows that some methods which are successful in RMSE based metrics and functional view of recommendation need not be successful in measure Next-k, 1-hit.

Main task was to discuss our results distributed along users – this gives more insight than global averages of user's achievements and also challenges future research. Similarly as in [10], we define transparency/difficulty of a user by the number of methods where at least 1-hit into target set was achieved.

Acknowledgment. This work was supported by Czech grants P103-15-19877S and P46.

References

1. Amatriain, X., Pujol, J.M., Oliver, N.: I like it… i like it not: evaluating user ratings noise in recommender systems. In: Houben, G.-J., McCalla, G., Pianesi, F., Zancanaro, M. (eds.) UMAP 2009. LNCS, vol. 5535, pp. 247–258. Springer, Heidelberg (2009)
2. Chapman, P.: The CRISP-DM user guide. In: 4th CRISP-DM SIG Workshop in Brussels in March 1999 (1999). http://en.wikipedia.org/wiki/Cross_Industry_Standard_Process_for_Data_Mining
3. Guillou, F., Gaudel, R., Mary, J., Preux, P.: User engagement as evaluation: a ranking or a regression problem? In: ACM 2014, pp. 7–12 (2014)
4. Horváth, T., Vojtáš, P.: Induction of fuzzy and annotated logic programs. In: Muggleton, S. H., Otero, R., Tamaddoni-Nezhad, A. (eds.) ILP 2006. LNCS (LNAI), vol. 4455, pp. 260–274. Springer, Heidelberg (2007)
5. Kifer, M., Subrahmanian, V.S.: Theory of generalized annotated logic programming and its applications. J. Logic Programm. **12**(4), 335–367 (1992)
6. Kopecky, M., Peska, L., Vojtas, P., Vomlelova, M.: Monotonization of user preferences. In: Andreasen, T., et al. (eds.) FQAS 2015. AISC, vol. 400, pp. 29–40. Springer, Heidelberg (2015)
7. Peska, L., Vojtas, P.: Hybrid recommending exploiting multiple DBPedia language editions. In: ESWC 2014 Linked Open Data-enabled Recommender Systems Challenge (2014)
8. Peska, L., Vojtas, P.: Hybrid biased k-NN to predict movie tweets popularity, poster. http://2014.recsyschallenge.com/program/SemWexMFF_short_09-21.pdf
9. RecSysChallenge2014: http://2014.recsyschallenge.com/leaderboard/, http://2014.recsys-challenge.com/
10. Smith-Miles, K.: Understanding strengths and weaknesses of optimization algorithms with new visualization tools and methodologies. In: IFORS Conference Plenary Session, Barcelona, July 2014. http://www.ifors2014.org/files2/KateSmithMiles.pdf
11. Vomlelova, M., Kopecky, M., Vojtas, P.: Transformation and aggregation preprocessing for top-k recommendation GAP rules induction. In: CEUR Proceedings Vol-1417 Rule Challenge and Doctoral Consortium @ RuleML 2015, Track 2: Rule-based Recommender Systems for the Web of Data. http://ceur-ws.org/Vol-1417/paper18.pdf

12. Kuchar, J.: Augmenting a feature set of movies using linked open data. In: CEUR Proceedings Vol-1417 Rule Challenge and Doctoral Consortium @ RuleML 2015, Track 2: Rule-based Recommender Systems for the Web of Data. http://ceur-ws.org/Vol-1417/paper16.pdf, Dataset http://nbviewer.ipython.org/urls/s3-eu-west-1.amazonaws.com/recsysrules2015/ RecSysRules2015-Dataset.ipynb

Span-Program-Based Quantum Algorithms for Graph Bipartiteness and Connectivity

Agnis Āriņš[✉]

University of Latvia, Raiņa Bulvāris 19, Riga 1586, Latvia
agnis.arins@lu.lv

Abstract. Span program is a linear-algebraic model of computation which can be used to design quantum algorithms. For any Boolean function there exists a span program that leads to a quantum algorithm with optimal quantum query complexity. In general, finding such span programs is not an easy task.

In this work, given a query access to the adjacency matrix of a simple graph G with n vertices, we provide two new span-program-based quantum algorithms:

- an algorithm for testing if the graph is bipartite that uses $O(n\sqrt{n})$ quantum queries;
- an algorithm for testing if the graph is connected that uses $O(n\sqrt{n})$ quantum queries.

1 Introduction

The concept of a span program as a linear-algebraic model of computation is not new. It was introduced by Karchmer and Wigderson in 1993 [9] and has many applications in classical complexity theory. Span programs can be used to evaluate decision problems. In 2008 Reichardt and Spalek [12] introduced a new complexity measure for span programs – witness size, which, as Reichardt showed later in [10,11], has strong connection with the quantum query complexity. There is a quantum algorithm for evaluating span programs [10] and these two complexity measures are essentially equivalent. The difficulty is to come up with a span program with a good witness size complexity.

In [12] authors dealt with bounded-size span programs evaluating Boolean functions each on $O(1)$ bits and posed an open question – do there exist interesting quantum algorithms based directly on asymptotically large span program? Belovs used span programs to construct learning graphs [3,4]. He also used span program approach for the matrix rank problem [2]. In [1] Ambainis et al. came up with a simple yet powerful span program for the graph collision problem.

In this paper, we extend the family of algorithms based on span programs. We present two new span-program-based quantum algorithms – an $O(n\sqrt{n})$ algorithm for the graph bipartiteness problem and an $O(n\sqrt{n})$ algorithm for the

This work has been supported by the ERC ADVANCED GRANT Methods for Quantum Computing.

© Springer International Publishing Switzerland 2016
J. Kofroň and T. Vojnar (Eds.): MEMICS 2015, LNCS 9548, pp. 35–41, 2016.
DOI: 10.1007/978-3-319-29817-7_4

graph connectivity problem. Both algorithms in the quantum query sense are optimal because the witness sizes match the quantum query complexity lower bounds [6,7] for these problems. Thus we demonstrate that span programs can be useful also for the problems with an asymptotically large input and possibly our algorithms could be building blocks for bigger span programs in the future.

The graph connectivity problem has been studied before [7] and there already exists a $O(n\sqrt{n})$ quantum query algorithm which requires $O(n)$ qubits of quantum memory. The advantage of our algorithm is that it uses only $O(\log n)$ qubits of quantum memory because the span program P_2 uses a vector space with $O(n^2)$ dimensions. Similarly for the graph bipartiteness problem. It can be solved with the breadth-first search method [8] which uses $O(n)$ qubits of quantum memory, but our approach with a span program requires $O(\log n)$ qubits of quantum memory.

2   Preliminaries

In this paper, we present algorithms which work on simple graphs, given in adjacency model. If the given graph has n vertices then the input size for an algorithm is $n \times n$ and we assume that the input variable $x_{i,j}$ corresponds to the value of entry in i-th row and j-th column of the adjacency matrix.

2.1   Span Programs

Definition 1 [1]. *A* span program *P is a tuple $P = (H, |t\rangle, V)$, where H is a finite-dimensional Hilbert space, $|t\rangle \in H$ is called the* target vector, *and $V = \{V_{i,b} | i \in [n], b \in \{0,1\}\}$, where each $V_{i,b} \subseteq H$ is a finite set of vectors.*

Denote by $V(x) = \bigcup\{V_{i,b} | i \in [n], x_i = b\}$. The span program is said to compute function $f : D \rightarrow \{0,1\}$, where the domain $D \subseteq \{0,1\}^n$, if for all $x \in D$,

$$f(x) = 1 \iff |t\rangle \in \mathrm{span}(V(x)).$$

Basically, what this definition says is that for each input variable x_i we have two sets of vectors (as the span program authors, we define these vectors in advance) – $V_{i,0}$ and $V_{i,1}$. If $x_i = b$ then we say that vectors from the set $V_{i,b}$ are available and vectors from the set $V_{i,1-b}$ are not available. If some vector is included in both sets $V_{i,0}$ and $V_{i,1}$ then we say that it is a free vector – it is always available.

The function returns 1 iff the target vector can be expressed as a linear combination of the available vectors, otherwise it returns 0.

Definition 2 [1].

(1) A positive witness *for $x \in f^{-1}(1)$ is a vector $w = (w_v), v \in V(x)$, such that $|t\rangle = \sum_{v \in V(x)} w_v v$. The* positive witness size *is*

$$\mathrm{wsize}_1(P) := \max_{x \in f^{-1}(1)} \min_{w : witness\ of\ x} \|w\|^2.$$

(2) *A* negative witness *for* $x \in f^{-1}(0)$ *is a vector* $w \in H$, *such that* $\langle t|w \rangle = 1$ *and for all* $v \in V(x)$: $\langle v|w \rangle = 0$. *The* negative witness size *is*

$$\text{wsize}_0(P) := \max_{x \in f^{-1}(0)} \min_{w:witness \ of \ x} \sum_{v \in V} \langle v|w \rangle^2.$$

(3) *The* witness size of a program P *is*

$$\text{wsize}(P) := \sqrt{\text{wsize}_0(P) \cdot \text{wsize}_1(P)}.$$

(4) *The* witness size of a function f *denoted by* $\text{wsize}(f)$ *is the minimum witness size of a span program that computes* f.

Theorem 1 [10,11]. $Q(f)$ *and* $\text{wsize}(f)$ *coincide up to a constant factor. That is, there exists a constant* $c > 1$ *which does not depend on* n *or* f *such that* $\frac{1}{c} \text{wsize}(f) \leq Q(f) \leq c \cdot \text{wsize}(f)$.

3   Span Program for Testing Graph Bipartiteness

A bipartite graph is a graph whose vertices can be divided into two disjoint sets such that there is no edge that connects vertices of the same set. An undirected graph is bipartite iff it has no odd cycles.

Algorithm 1. *There exists a span program* P *which for a graph with* n *vertices detects if the graph is bipartite with* $\text{wsize}(P) = O(n\sqrt{n})$.

Proof. We will make a span program which detects if a graph has an odd cycle.
     Let $n = |G|$ be a number of vertices in the given graph G. Then the span program is as follows:

Span program P_1 for testing graph bipartiteness

- H is a $(2n^2 + 1)$ dimensional vector space with basis vectors $\{|0\rangle\} \cup \{|v_{k,b}\rangle \ |v, k \in [1..n], b \in \{0,1\}\}$.
- The target vector is $|0\rangle$.
- For every $k \in [1..n]$ make available the free vector $|0\rangle + |k_{k,0}\rangle + |k_{k,1}\rangle$.
- For every $k \in [1..n]$, for every edge $u - v$ (where input $x_{u,v} = 1$), make available the vectors $|u_{k,0}\rangle + |v_{k,1}\rangle$ and $|u_{k,1}\rangle + |v_{k,0}\rangle$.

The states in the span program P_1 are mostly in the form $|v_{k,b}\rangle$ where v is vertex index, k represents vertex from which we started our search for an odd length cycle and b represents the parity of the current path length. The first subindex k in state $|v_{k,b}\rangle$ can also be considered as the subspace index for the subspace $V_k = \text{span}(\{|v_{k,b}\rangle \ |v \in [1..n], b \in \{0,1\}\})$. Vectors corresponding to edges are in the form $|u_{k,b}\rangle + |v_{k,1-b}\rangle$ consisting from sum of two states which both belong to same subspace V_k.

In the span program P_1 the target vector $|0\rangle$ can only be expressed as a linear combination of the available vectors if at least one of the vectors in the form $|k_{k,0}\rangle + |k_{k,1}\rangle$ can be expressed. Without loss of generality, if there is an odd length cycle $v_1 - v_2 - \cdots - v_{(2j+1)} - v_1$ then the target vector can be expressed by taking the vectors corresponding to the edges of this cycle, alternatingly with plus and minus sign

$$|0\rangle = (|0\rangle + |1_{1,0}\rangle + |1_{1,1}\rangle) - (|1_{1,0}\rangle + |2_{1,1}\rangle) + \cdots - (|(2j+1)_{1,0}\rangle + |1_{1,1}\rangle)$$

therefore the span program P_1 will always return 1 when the given graph is not bipartite.

From the other side, if there is no odd length cycle then none of the vectors in the form of $|k_{k,0}\rangle + |k_{k,1}\rangle$ can be expressed using the available vectors from P_1. To cancel out the state $|k_{k,0}\rangle$ we should be using a vector $|k_{k,0}\rangle + |v_{k,1}\rangle$ corresponding to some edge $k - v$ where v is some vertex adjacent to k because no other vector contains the state $|k_{k,0}\rangle$. By doing so we move from the state $|k_{k,0}\rangle$ to the state $|v_{k,1}\rangle$ (possibly with some coefficient other than 1) which has the parity bit flipped. Similarly, to cancel out the state $|v_{k,1}\rangle$ we should be using a vector corresponding to some edge going out from vertex v. To stop this process we need to reach the state $|k_{k,1}\rangle$. It can be done only if there is an odd cycle because the path must be closed and the parity bit restricts it to odd length. When there is no odd cycle, span program P_1 will always return 0.

We can conclude that P_1 indeed computes the expected function. It remains to calculate the witness size of P_1.

For the case when there is an odd cycle we need to calculate the positive witness size. If there is an odd cycle $v_1 - v_2 - \cdots - v_d - v_1$ with length d then the target vector can be expressed in this way

$$|0\rangle = 1 \cdot (|0\rangle + |1_{1,0}\rangle + |1_{1,1}\rangle) + (-1) \cdot (|1_{1,0}\rangle + |2_{1,1}\rangle) + \cdots + (-1) \cdot (|d_{1,0}\rangle + |1_{1,1}\rangle)$$

and the positive witness w here consists only from $d + 1$ entries ± 1 therefore $\|w\|^2 = d + 1$.

If $v_1 - v_2 - \cdots - v_d - v_1$ is a cycle then also $v_2 - v_3 - \cdots - v_d - v_1 - v_2$ is a cycle and therefore the target vector can also be expressed in this way

$$|0\rangle = (|0\rangle + |2_{2,0}\rangle + |2_{2,1}\rangle) - (|2_{2,0}\rangle + |3_{2,1}\rangle) + \cdots - (|1_{2,0}\rangle + |2_{2,1}\rangle)$$

the same follows for all d vertices in this cycle and the target vector therefore can be expressed in atleast d different ways. We can combine these d ways each taken with coefficient $1/d$ and then we get that the positive witness size

$$\text{wsize}_1(P_1) \leq d * (1/d)^2 * (d + 1) < 2 \tag{1}$$

To estimate the negative witness size we must find a negative witness w'. We derive w' by defining how it acts on basis vectors. From definition $\langle w'|0\rangle = 1$.

For every k we must have $\langle w'| (|0\rangle + |k_{k,0}\rangle + |k_{k,1}\rangle)) = 0$ therefore lets pick w' in such a way that $\langle w'|k_{k,0}\rangle = 0$ and $\langle w'|k_{k,1}\rangle = -1$. Now repeat the following steps until no changes happen:

- for every available vector $|u_{k,0}\rangle + |v_{k,1}\rangle$ if $\langle w'|u_{k,0}\rangle$ is defined then define $\langle w'|v_{k,1}\rangle = -\langle w'|u_{k,0}\rangle$.
- for every available vector $|u_{k,1}\rangle + |v_{k,0}\rangle$ if $\langle w'|u_{k,1}\rangle$ is defined then define $\langle w'|v_{k,0}\rangle = -\langle w'|u_{k,1}\rangle$.

For all not yet defined $\langle w'|v_{k,b}\rangle$ define $\langle w'|v_{k,b}\rangle = 0$.

For any given vector v in span program P_1 the value $\langle w'|v\rangle^2 \le 1$. The total number of vectors does not exceed $n + n^3$ therefore the negative witness size is

$$\text{wsize}_0(P_1) \le 1 \cdot (n + n^3) \tag{2}$$

Combining positive and negative witness sizes we obtain the upper bound for witness size which also corresponds to quantum query complexity

$$\text{wsize}(P_1) = \sqrt{\text{wsize}_0(P_1) \cdot \text{wsize}_1(P_1)} = O\left(n\sqrt{n}\right) \tag{3}$$

$\square$

4 Span Program for Testing Graph Connectivity

A graph is said to be connected if every pair of vertices in the graph is connected. If in an undirected graph one vertex is connected to all other vertices then by transitivity the graph is connected.

Algorithm 2. *There exists a span program P which for a graph with n vertices detects if the graph is connected with $wsize(P) = O(n\sqrt{n})$.*

Proof. Let $n = |G|$ be a number of vertices in the given graph G. Then the span program is as follows:

Span program P_2 for testing graph connectivity

- H is a $n^2 - 1$ dimensional vector space with basis vectors $\{|v_k\rangle \,|\, v \in [0..n], k \in [2..n]\}$.
- The target vector is $|t\rangle = |0_2\rangle + |0_3\rangle + \cdots + |0_n\rangle$.
- For every $k \in [2..n]$ make available the free vector $|0_k\rangle + |1_k\rangle - |k_k\rangle$.
- For every $k \in [2..n]$, for every edge $u - v$ (where $u, v \in [1..n]$ and input $x_{u,v} = 1$), make available the vector $|u_k\rangle - |v_k\rangle$.

If all vertices are reachable from vertex with index 1 then the given graph is connected. Here we use Belov's [5] span program for *s-t connectivity* as subroutine. This subroutine checks if in a given graph there is a path from the vertex s to the vertex t. The span program for it has the target vector $|s\rangle - |t\rangle$ and for each edge $i - j$ (input $x_{i,j} = 1$) we can use the vector $|i\rangle - |j\rangle$.

In span program P_2, by using this subroutine $n-1$ times, we check if all other vertices are connected to vertex with index 1. We create a separate subspace $V_k = \text{span}(\{|v_k\rangle \,|\, v \in [0..n]\})$ for each such subroutine call to avoid any interference between them, which is a common technique [10] how to compose span programs. The span program returns 1 when all vertices are connected, but otherwise it returns 0.

For the case when the given graph is connected we need to calculate the positive witness size. In each *s-t* subroutine the shortest path length from the vertex s to the vertex t can not be larger than $n - 1$. Therefore each vector from the set $\{|0_k\rangle \,|\, k \in [2..n]\}$ requires no more than n vectors to express it. There are $n - 1$ such subroutines. The positive witness size is

$$wsize_1(P_2) \leq n \cdot (n-1) \leq n^2 \tag{4}$$

To estimate the negative witness size we must find a negative witness w'. We derive w' by defining how it acts on the basis vectors. From definition $\langle w'|t\rangle = 1$. We need to talk about negative witness only when some vertex v is not connected to vertex with index 1. Then the vertex v belongs to different connected component than vertex with index 1. Lets name this connected component C_v and let the count of vertices in this connected component be d_v. Lets pick w' in such a way that for each vertex $v_k \in C_v$ set $\langle w'|0_k\rangle = 1/d_v$ and for each vertex $v_z \notin C_v$ set $\langle w'|0_z\rangle = 0$.

For $k \in [2..n]$ we must have $\langle w'|(|0_k\rangle + |1_k\rangle - |k_k\rangle)\rangle = 0$ therefore set $\langle w'|1_k\rangle = -\langle w'|0_k\rangle$ and $\langle w'|k_k\rangle = 0$. Now repeat the following step until no changes happen: for every available vector $|u_k\rangle - |v_k\rangle$ if $\langle w'|u_k\rangle$ is defined then define $\langle w'|v_k\rangle = \langle w'|u_k\rangle$. For all other not yet defined basis vectors $|v_k\rangle$ set $\langle w'|v_k\rangle = 0$.

With such negative witness w' choice the overall negative witness size will only get increased by vectors which correspond to nonexistent graph edges which connects C_v with other connected components in graph - i.e. border edges. An edge $u-v$ is a border edge if $u \in C_v$ and $v \notin C_v$. To a border edge $u-v$ correspond the vectors $|u_k\rangle - |v_k\rangle$ where $k \in [2..n]$ but only d_v from these vectors will have $\langle w'|u_k\rangle \neq \langle w'|v_k\rangle$ and each such vector increases the negative witness size by value $(1/d_v)^2$. For C_v there are at most $d_v \cdot (n-1)$ border edges therefore the overall negative witness size is

$$wsize_0(P_2) \leq d_v^2 \cdot (n-1) \cdot (1/d_v)^2 \leq n \tag{5}$$

Combining the positive and negative witness sizes we obtain the upper bound for the witness size which also corresponds to the quantum query complexity

$$wsize(P_2) = \sqrt{wsize_0(P_2) \cdot wsize_1(P_2)} = O\left(n\sqrt{n}\right) \tag{6}$$

$\square$

Acknowledgements. I am grateful to Andris Ambainis for the suggestion to solve the graph problems with span programs, and for many useful comments during the development of the paper.

References

1. Ambainis, A., Balodis, K., Iraids, J., Ozols, R., Smotrovs, J.: Parameterized quantum query complexity of graph collision. CoRR abs/1305.1021 (2013). http://arxiv.org/abs/1305.1021
2. Belovs, A.: Span-program-based quantum algorithm for the rank problem. CoRR abs/1103.0842 (2011). http://arxiv.org/abs/1103.0842
3. Belovs, A.: Learning-graph-based quantum algorithm for k-distinctness. In: IEEE 53rd Annual Symposium on Foundations of Computer Science (FOCS), pp. 207–216. IEEE (2012). http://ieeexplore.ieee.org/xpls/abs_all.jsp?arnumber=6375298
4. Belovs, A.: Span programs for functions with constant-sized 1-certificates. In: Proceedings of the 44th Symposium on Theory of Computing, pp. 77–84. ACM (2012). http://dl.acm.org/citation.cfm?id=2213985
5. Belovs, A., Reichardt, B.W.: Span programs and quantum algorithms for st-connectivity and claw detection. In: Epstein, L., Ferragina, P. (eds.) ESA 2012. LNCS, vol. 7501, pp. 193–204. Springer, Heidelberg (2012). http://dx.doi.org/10.1007/978-3-642-33090-2_18
6. Berzina, A., Dubrovsky, A., Freivalds, R., Lace, L., Scegulnaja, O.: Quantum query complexity for some graph problems. In: Van Emde Boas, P., Pokorný, J., Bieliková, M., Štuller, J. (eds.) SOFSEM 2004. LNCS, vol. 2932, pp. 140–150. Springer, Heidelberg (2004). http://dx.doi.org/10.1007/978-3-540-24618-3_11
7. Dürr, C., Heiligman, M., Høyer, P., Mhalla, M.: Quantum query complexity of some graph problems. In: Díaz, J., Karhumäki, J., Lepistö, A., Sannella, D. (eds.) ICALP 2004. LNCS, vol. 3142, pp. 481–493. Springer, Heidelberg (2004). http://dx.doi.org/10.1007/978-3-540-27836-8_42
8. Furrow, B.: A panoply of quantum algorithms. Quantum Info. Comput. 8(8), 834–859 (2008). http://dl.acm.org/citation.cfm?id=2017011.2017022
9. Karchmer, M., Wigderson, A.: On span programs. In: Proceedings of the Eighth Annual Structure in Complexity Theory Conference, pp. 102–111. IEEE (1993). http://ieeexplore.ieee.org/xpls/abs_all.jsp?arnumber=336536
10. Reichardt, B.W.: Span programs and quantum query complexity: the general adversary bound is nearly tight for every boolean function. In: 50th Annual IEEE Symposium on Foundations of Computer Science (FOCS), pp. 544–551. IEEE (2009). http://ieeexplore.ieee.org/xpls/abs_all.jsp?arnumber=5438598
11. Reichardt, B.W.: Reflections for quantum query algorithms. In: Proceedings of the Twenty-Second Annual ACM-SIAM Symposium on Discrete Algorithms, pp. 560–569. SIAM (2011). http://dl.acm.org/citation.cfm?id=2133080
12. Reichardt, B.W., Spalek, R.: Span-program-based quantum algorithm for evaluating formulas. In: Proceedings of the Fortieth Annual ACM Symposium on Theory of Computing, STOC 2008, pp. 103–112. ACM, New York (2008). http://doi.acm.org/10.1145/1374376.1374394

Fitting Aggregation Operators

Vojtěch Havlena[1][(⊠)] and Dana Hliněná[2]

[1] Faculty of Information Technology, Brno University of Technology,
Brno, Czech Republic
xhavle03@stud.fit.vutbr.cz
[2] Faculty of Electrical Engineering and Communication,
Brno University of Technology, Brno, Czech Republic
hlinena@feec.vutbr.cz

Abstract. This paper treats the problem of fitting aggregation operators to empirical data. Specifically, we are interested in modelling of the conjunction in human language. To our knowledge, the first attempt to see how humans "interpret" the conjunction for graded properties is due to the paper [1]. In that case, simply the minimum t-norm came out. Our results are different because our approach to the resolution is different. We have experimentally rated simple statements and their conjunctions. Then we have tried, on the basis of measured data, to find a suitable function, which corresponds to human conjunction. First, we discuss methods applicable to associative operators, t-norms. Next, we propose an algorithm for approximation of the t-norm's generator based on the weighting method and Lawson-Hanson's algorithm. Suitable modifications of the algorithm can generalize our solutions to aggregation operators. In this way we get new results for generated means which are well-known representatives of aggregation operators. Empirically measured data suggest that people do not understand conjunction necessarily as a commutative operation. Finally, we investigate the modelling of the conjunction via generated Choquet integral.

1 Preliminaries

To start, we repeat some important definitions and theorems. In fuzzy logic, conjunctions are often interpreted by the triangular norms.

Definition 1. [2] *A triangular norm (t-norm for short) is a binary operation on the unit interval $[0, 1]$, i.e., a function $T : [0, 1]^2 \to [0, 1]$, such that for all $x, y, z \in [0, 1]$ the following four axioms are satisfied:*

– *(T1) Commutativity*
$$T(x, y) = T(y, x),$$

– *(T2) Associativity*
$$T(T(x, y), z) = T(x, T(y, z)),$$

– *(T3) Monotonicity*
$$T(x, y) \leq T(x, z) \text{ whenever } y \leq z,$$

© Springer International Publishing Switzerland 2016
J. Kofroň and T. Vojnar (Eds.): MEMICS 2015, LNCS 9548, pp. 42–53, 2016.
DOI: 10.1007/978-3-319-29817-7_5

– *(T4) Boundary Condition*
$$T(x, 1) = x,$$

The four basic t-norms are:

- the minimum t-norm $T_M(x, y) = \min\{x, y\}$,
- the product t-norm $T_P(x, y) = x \times y$,
- the Łukasiewicz t-norm $T_L(x, y) = \max\{0, x + y - 1\}$,
- the drastic product $T_D(x, y) = \begin{cases} 0 & \text{if } \max\{x, y\} < 1, \\ \min\{x, y\} & \text{otherwise.} \end{cases}$

We will deal only with such continuous t-norms, that are generated by a unary function (generator). One possibility is to generate by an additive generator, which is a strictly decreasing function f from the unit interval $[0, 1]$ to $[0, +\infty]$ such that $f(1) = 0$ and $f(x) + f(y) \in H(f) \cup [f(0^+), +\infty]$ for all $x, y \in [0, 1]$. Then the generated t-norm is given as follows

$$T(x, y) = f^{(-1)}\left(f(x) + f(y)\right),$$

where $f^{(-1)} : [0, +\infty] \to [0, 1]$ and $f^{(-1)}(x) = \sup\{x \in [0, 1]; f(x) > y\}$. Note, that $f^{(-1)}$ is a pseudo-inverse, which is a monotone extension of the ordinary inverse function. For an illustration, we give the following examples of parametric classes of t-norms and their additive generators.

The family of Frank t-norms, introduced by M. J. Frank in the late 1970s, is given by the parametric definition for $0 \leq p \leq +\infty$ as follows:

$$T_p^F(x, y) = \begin{cases} T_M(x, y) & \text{if } p = 0, \\ T_P(x, y) & \text{if } p = 1, \\ T_L(x, y) & \text{if } p = +\infty, \\ \log_p\left(1 + \frac{(p^x - 1)(p^y - 1)}{p - 1}\right) & \text{otherwise.} \end{cases}$$

An additive generator for T_p^F is

$$f_p^F(x) = \begin{cases} -\log x & \text{if } p = 1, \\ 1 - x & \text{if } p = +\infty, \\ \log_p \frac{p - 1}{p^x - 1} & \text{otherwise.} \end{cases}$$

The family of Yager t-norms, introduced in the early 1980s by Ronald R. Yager, is given for $0 \leq p \leq +\infty$ by

$$T_p^Y(x, y) = \begin{cases} T_D(x, y) & \text{if } p = 0, \\ T_M(x, y) & \text{if } p = +\infty, \\ \max\left\{0, 1 - ((1 - x)^p + (1 - y)^p)^{\frac{1}{p}}\right\} & \text{if } 0 < p < +\infty. \end{cases}$$

The additive generator of T_p^Y for $0 < p < +\infty$ is

$$f_p^Y(x) = (1 - x)^p.$$

Because of associativity, we can extend t-norms to the n-variete case as:

$$x_T^{(n)} = \begin{cases} x & \text{if } n = 1, \\ T(x, x_T^{(n-1)}) & \text{if } n > 1. \end{cases}$$

A t-norm T is called Archimedean if for each x, y in the open interval $]0, 1[$ there is a natural number n such that $x_T^{(n)} \leq y$. It is sufficient to investigate Archimedean t-norms, because every non-Archimedean t-norm can be approximated arbitrarily well with Archimedean t-norms [3,4]. For continuous Archimedean t-norms there exist additive generators $g : [0, 1] \rightarrow [0, +\infty]$, such that

$$T(x_1, x_2, \ldots, x_n) = g^{(-1)}\left(\sum_{i=1}^{n} g(x_i)\right).$$

The generator g is strictly monotone decreasing with $g(1) = 0$ and either $g(0) = +\infty$ or $g(0) = a < +\infty$.

One of the generalizations of t-norms are aggregation operators A_n, which are monotone increasing functions $A_n : [0, 1]^n \rightarrow [0, 1]$ with the boundary conditions: $A_n(0, 0, \ldots, 0) = 0, A_n(1, 1, \ldots, 1) = 1$. The most popular aggregation operators are t-norms, t-conorms, uninorms, generalized means and ordered weighted aggregation operators. In this paper we deal with generated quasi-arithmetic means which are given by

$$M(x_1, x_2, \ldots, x_n) = g^{(-1)}\left(\sum_{i=1}^{n} \frac{1}{n} g(x_i)\right),$$

where g is a monotone increasing function $[0, 1] \rightarrow [0, 1]$.

At the conclusion, we will work with Choquet integral as the aggregation operator, so we repeat the definitions of universal fuzzy measures and Choquet integrals:

Definition 2. [5] *Let* $N = \{1, 2, \ldots, m\}$ *and* $\mathcal{A} = \{(n, E); n \in N, E \subseteq N\}$. *A mapping* $M : \mathcal{A} \rightarrow [0, 1]$ *is called a universal fuzzy measure whenever for each fixed* $n \in N, M(n, .)$ *is a fuzzy measure, that is*

– $M(n, \emptyset) = 0, M(n, N) = 1$,
– $M(n, E) \leq M(n, F)$ *for all* $E \subseteq F \subseteq N$.

For a given universal fuzzy measure M, an aggregation operator can be constructed by means of any fuzzy integral. We turn our attention to the Choquet integral and we get

$$A(x_1, x_2, \ldots, x_n) = (C)\int_N x \, dm_n = \sum_{i=1}^{n} (x_{\alpha(i)} - x_{\alpha(i-1)}) M(n, E_{\alpha(i)}),$$

where $m_n = M(n, .)$, α is a permutation of $(1, 2, \ldots, n)$ yielding $x_{\alpha(1)} \le x_{\alpha(2)} \le \cdots \le x_{\alpha(n)}$, $E_{\alpha(i)} = \{\alpha(i), \ldots, \alpha(n)\}$ and $x_{\alpha(0)} = 0$ by convention. The corresponding Choquet integral based aggregation operator is given by

$$A^g(x_1, x_2, \ldots, x_n) = \sum_{i=1}^{n} x_{\alpha(i)} \left(g\left(\frac{i}{n}\right) - g\left(\frac{i-1}{n}\right) \right) = \sum_{i=1}^{n} w_{i,n} x_{\alpha(i)}.$$

2 Generator's Approximation Algorithm

Beliakov and others in [5,6] introduced additive generator's approximation of some aggregation operators from empirical data via B-splines. The generator is represented by

$$g(x) = S_{m,t}(x) = \sum_{j=1}^{J} c_j B_{j,m}(x),$$

where $\mathbf{c}$ is the vector of coefficients and $B_{j,m}(x)$ are B-spline basic functions of order m. These functions are defined by the following recurrent formula [7]

$$B_{i,1}(x) = \begin{cases} 1 & \text{if } t_i \le x < t_{i+1}, \\ 0 & \text{otherwise}, \end{cases}$$

$$B_{i,n}(x) = \frac{x - t_i}{t_{i+n-1} - t_i} B_{i,n-1}(x) + \frac{t_{i+n} - x}{t_{i+n} - t_{i+1}} B_{i+1,n-1}(x),$$

where $t_1, \ldots, t_{J+m}$ is a non-decreasing sequence of real numbers called the knot sequence. Using the definitions from the previous section and generator's boundary conditions, the approximation problem is given as follows

$$\mathbf{Ac} \approx \mathbf{b}, \quad \mathbf{Dc} \ge \mathbf{0}, \quad \mathbf{Ec} = \mathbf{d}, \tag{1}$$

The first system determines the shape of an additive generator with respect to empirical data. Matrices $\mathbf{A}, \mathbf{b}$ are given by additive generator's representation of aggregation operators. Their exact values then depend on empirical data. Vector $\mathbf{c}$ is the unknown vector of B-spline coefficients. Matrices $\mathbf{E}, \mathbf{d}$ are given by generator's boundary conditions and $\mathbf{D}$ is either $-\mathbf{I}$ or $\mathbf{I}$. The exact form of these matrices depends on a concrete aggregation operator and it is described in detail in [5]. This paper deals with an algorithmic solution for the approximation. If the conditions (1) did not contain constrains on inequality and equality, the generator's approximation problem would be easily solvable by the method of the least squares. The mentioned method solves the following approximation problem [8]

$$\mathbf{Ax} \approx \mathbf{b}, \quad \text{where } \mathbf{A} \in \mathbb{R}^{n \times m}, \ \mathbf{x} \in \mathbb{R}^m, \ \mathbf{b} \in \mathbb{R}^n. \tag{2}$$

The resulting vector $\mathbf{x}_{LE}$ is given by the expression

$$\mathbf{x}_{LE} = \arg\min_{\mathbf{x}} \|\mathbf{Ax} - \mathbf{b}\|_2^2.$$

However, due to our constrains, we have to use a more sophisticated method than the classical least squares method. Since we have two restriction conditions, we can divide algorithm into independent parts. Each of them will be explained in the following text.

2.1   Method of Weighting

The first part is devoted to the equality $\mathbf{Ex} = \mathbf{d}$. The method of direct elimination or the method of weighting can be used for solving this approximation problem. In our case the method of weighting is chosen because of its simplicity. The method is based on an assignment of sufficiently big weights to the restricted coefficients. This approach is summarized by following identity [8]

$$\mathbf{x}_{LSE} = \arg\min_{\mathbf{x}} \lim_{\gamma \to \infty} \left\| \begin{pmatrix} \gamma\mathbf{E} \\ \mathbf{A} \end{pmatrix} \mathbf{x} - \begin{pmatrix} \gamma\mathbf{d} \\ \mathbf{b} \end{pmatrix} \right\|_2^2. \tag{3}$$

where $\mathbf{x}_{LSE}$ is the optimal solution of the least squares. It is evident from this formula that the problem with the equality can be converted to the problem solvable by ordinary least squares. Note that value γ is chosen to be significantly bigger than the biggest value of the matrix $\mathbf{A}$.

2.2   Lawson-Hanson's Algorithm

In the previous section we have shown the method of weighting that solves the equality. Now we will investigate the inequality, more precisely nonnegativity. Lawson-Hanson's algorithm allows us to solve such restricted optimization problems, based on least squares [9]. This algorithm looks for a vector $\mathbf{x}$ fulfilling the following condition

$$\min_{\mathbf{x}} \|\mathbf{Ax} - \mathbf{b}\|_2^2, \text{ subject to } \mathbf{x} \geq 0.$$

The algorithm is based on the active set method. This set includes indices of the variables whose regression coefficients are negative or zero. The remaining indices of the variables are included to the passive set. If the passive set and the active set are known, such restricted optimization problem's solution is obtained by the least squares method with variables in the passive set. Coefficients of variables in the active set are set to zero.

Lawson-Hanson's algorithm is an iterating algorithm. Current passive set is determined in every iteration. The passive and the active sets are modified according to results of the least squares over variables in the passive set. The modification consists of exchanging the variables between each set. Simultaneously a new value of vector $\mathbf{x}$ is computed. During the set modification, indices with zero value in vector $\mathbf{x}$ are exchanged. The algorithm in pseudo-code is shown in Algorithm 1 [9].

Note that the passive set is denoted as P, the active set as R. The number of iterations can be affected by the parameter *tolerance*. Notation $\mathbf{A}^P$ denotes matrix associated with the indexes of variables in the passive set P. The least squares are directly computed on lines 8 and 13.

Algorithm 1. LAWSON-HANSON'S

Input: $\mathbf{A} \in \mathbb{R}^{m \times n}, \mathbf{b} \in \mathbb{R}^m$
Output: $\mathbf{x}^* \geq 0$, where $\mathbf{x}^* = \arg\min \|\mathbf{A}\mathbf{x} - \mathbf{b}\|_2^2$

1: $P \leftarrow \emptyset$
2: $R \leftarrow \{1, 2, \ldots, n\}$
3: $\mathbf{x} \leftarrow \mathbf{0}$
4: $\mathbf{w} \leftarrow \mathbf{A}^T(\mathbf{y} - \mathbf{A}\mathbf{x})$
5: **while** $R \neq \emptyset \wedge \max_{i \in R}(w_i) > tolerance$ **do**
6:     $j \leftarrow \arg\max_{i \in R}(w_i)$
7:     Include index j into passive set P and remove j from R
8:     $\mathbf{s}^P \leftarrow [(\mathbf{A}^P)^T \mathbf{A}^P]^{-1} (\mathbf{A}^P)^T \mathbf{b}$
9:     **while** $\min(\mathbf{s}^P) \leq 0$ **do**
10:        $\alpha \leftarrow -\min_{i \in P}(x_i/(x_i - s_i))$
11:        $\mathbf{x} \leftarrow \mathbf{x} + \alpha(\mathbf{s} - \mathbf{x})$
12:        Move from P to R all indexes i such as $x_i = 0$
13:        $\mathbf{s}^P \leftarrow [(\mathbf{A}^P)^T \mathbf{A}^P]^{-1} (\mathbf{A}^P)^T \mathbf{b}$
14:        $\mathbf{s}^R \leftarrow \mathbf{0}$
15:     **end while**
16:     $\mathbf{x} \leftarrow \mathbf{s}$
17:     $\mathbf{w} \leftarrow \mathbf{A}^T(\mathbf{y} - \mathbf{A}\mathbf{x})$
18: **end while**
19: **return x**

2.3 T-norm's Generator Approximation Algorithm

As mentioned before, Lawson-Hanson's algorithm outputs nonnegative coefficients. But in our case, we require a decreasing approximated additive generator. It is ensured by negative coefficients of the B-spline. The first step is thus a transformation of the input empirical data and boundary conditions. Transformation and inverse transformation are given as follows

$$\mathcal{F}(x) = \mathcal{F}^{-1}(x) = 1 - x. \tag{4}$$

Using such transformated data, we specify the matrices $\mathbf{A}$ and $\mathbf{E}$. The boundary conditions are guaranteed by the method of weighting. The coefficients of an approximated B-spline are computed by Lawson-Hanson's algorithm. Due to data transformation a resultant curve is increasing, so the inverse tranformation (4) of a curve values is performed. This step ensures decreasing approximated B-spline with demanded boundary conditions. Algorithm in pseudo-code is shown in Algorithm 2 [10].

Note that the method of weighting is used on line 7. Lawson-Hanson's algorithm is in pseudo-code represented by function *nonnegative*. The number of B-spline's nodes and its values depend on specific empirical data and a choice is made experimentally. The shape of the obtained curve is affected by these values.

Algorithm 2. APPROXIMATION OF THE ADDITIVE GENERATOR

Input: The set of empirical data in the form $\{\mathbf{x}^k, y^k\}, k = 1, 2, \dots, n$
Output: Values $\mathbf{y}$ of the approximated generator with constant step h

1: $\mathbf{A} \leftarrow \mathbf{0}, \mathbf{b} \leftarrow \mathbf{0}, \mathbf{C} \leftarrow \mathbf{0}, \mathbf{d} \leftarrow (0 \quad 1)^T$
2: $\widetilde{\mathbf{x}}^k \leftarrow \mathcal{F}(\mathbf{x}^k), \widetilde{y}^k \leftarrow \mathcal{F}(y^k)$ pro $k = 1, 2, \dots, n$
3: Choice of the B-spline's degree m and values ε and γ
4: Setting number of the B-spline's nodes J and its values $\mathbf{t}$
5: $A_{kj} \leftarrow \sum_{i=1}^{n_k} B_{j,m}(\widetilde{x}_i^k) - B_{j,m}(\widetilde{y}^k)$
6: $E_{0j} \leftarrow B_{m,j}(\mathcal{F}(1)), E_{1j} \leftarrow B_{m,j}(\mathcal{F}(\varepsilon))$
7: $\mathbf{A} \leftarrow \begin{pmatrix} \gamma\mathbf{E} \\ \mathbf{A} \end{pmatrix}, \mathbf{b} \leftarrow \begin{pmatrix} \gamma\mathbf{d} \\ \mathbf{b} \end{pmatrix}$
8: $\mathbf{c} \leftarrow nonnegative(\mathbf{A}, \mathbf{b})$
9: $i \leftarrow \varepsilon$
10: $i \leftarrow h$
11: **while** $i \leq 1$ **do**
12:      **if** $i < \varepsilon$ **then**
13:          $y(i) \leftarrow \frac{1}{i} + 1 - \frac{1}{\varepsilon}$
14:      **else**
15:          $y(i) \leftarrow \sum_{j=1}^{J} c_j B_{j,m}(\mathcal{F}^{-1}(i))$
16:      **end if**
17:      $0\ i \leftarrow i + h$
18: **end while**

To obtain a better notion about the algorithm, let's make an estimate of a time complexity. Lawson-Hanson's algorithm iteratively performs computation of least squares. The time complexity of least squares can be divided into following suboperations.

- The multiplication of $\mathbf{A}^T\mathbf{A}$ ($\mathbf{A} \in \mathbb{R}^{n \times m}, \mathbf{b} \in \mathbb{R}^m$) is performed with the complexity $\mathcal{O}(mn^2)$, $\mathbf{A}^T\mathbf{b}$ with complexity $\mathcal{O}(nm)$.
- Inverse matrix computation with $\mathcal{O}(n^3)$ [11].
- Multiplication of the obtained inverse matrix with complexity $\mathcal{O}(n^2)$.

Between the mentioned operations, the multiplication $\mathbf{A}^T\mathbf{A}$ and the inverse matrix computation dominate in the sense of time complexity. The resulting complexity estimate of least squares is thus $\mathcal{O}(n^2(n + m))$. The time complexity of Lawson-Hanson's algorithm depends, as mentioned before, on the number of iterations, at which the least squares are computed. Therefore the complexity is $\mathcal{O}(pn^2(n + m))$, where p is the maximum number of iterations. But this is only an estimate, the actual value is smaller, because the least squares are computed over matrix associated to passive set P ($|P| \leq n$).

Regarding the analysis of the entire algorithm, the weighting method is performed with time complexity $\mathcal{O}(Jn)$ and evaluating values of the B-spline with $\mathcal{O}(\lfloor\frac{1-\varepsilon}{h}\rfloor Jm)$. The most complex task is thus computing coefficients using Lawson-Hanson's algorithm. Hence the complexity of the entire algorithm is

$\mathcal{O}(pn^2(n+J))$. Since we assume that $J < n$, from the previous formula we get $\mathcal{O}(pn^3)$. Computing time can be improved by using specialized algorithms for matrix multiplication (e.g. Strassen's algorithm [11]) and inverse matrix computation [12] or using Cox-De Boor's algorithm for B-spline evaluation [13] (Fig. 1).

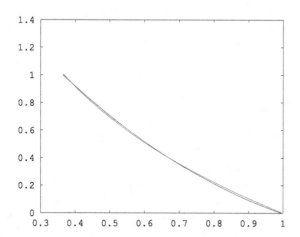

Fig. 1. Example of Product t-norm's additive generator approximation. Product t-norm's generator is given by $g(x) = -\log(x)$ in the interval $[e^{-1}, 1]$ (red curve). B-spline of degree 3 was chosen for approximation (blue curve). Coefficients computation was based on 8 empirical values with two arguments. Approximation error is 0.013 (Color figure online).

2.4   Algorithm Modifications

The algorithm of t-norm's generator approximation can be used with small modifications to approximate the generators of other classes of aggregation operators. In this paper, we consider means and Choquet integral based operators. In the case of means the generators are increasing and in the case of Choquet integral based operators the generators are non-decreasing. Due to this fact, transformation of the input data and boundary conditions is not performed (i.e. $\mathcal{F}(x) = \mathcal{F}^{-1}(x) = x$).

The boundary conditions are identical for both considered aggregation operators $(g(0) = 0, g(1) = 1)$. The ε value is thus set to zero. The 6th line of original algorithm is therefore modified as follows

$$E_{0j} = B_{m,j}(0), E_{1j} = B_{m,j}(1).$$

Computing of the matrix **A** is different in each case. For means it is given by the formula

$$A_{kj} = \frac{1}{n_k} \sum_{i=1}^{n_k} B_{j,m}(\widetilde{x}_i^k) - B_{j,m}(\widetilde{y}^k). \tag{5}$$

For Choquet integral based operator it is given by

$$A_{kj} = \sum_{i=1}^{n_k} \tilde{x}_i^k \left(B_{j,m}\left(\frac{i}{n_k}\right) - B_{j,m}\left(\frac{i-1}{n_k}\right)\right). \tag{6}$$

Note that the 5th line of original algorithm is replaced by (5) or (6). In the case of Choquet integral, initiation of vector $\mathbf{b}$ on the first line of original algorithm is replaced by $\mathbf{b} \leftarrow \mathbf{y}$, where $\mathbf{y}$ is the vector of empirical results. With the above mentioned modifications, the original algorithm can be generalized for approximation of certain class of aggregation operators. The time complexity of the modified algorithms corresponds to the original algorithm.

3   Collected Empirical Data

Empirical data for our experiment were obtained from the respondents by means of a paper questionnaire. The task of respondents was to assign a value in the range of 1–10 to each statement. This value represents a level of truthfulness according to respondent's opinion. There were 20 statements, 10 of them are elementary statements and the remaining 10 are composited to conjunction from the elementary ones. Each conjunction was included in the questionnaire in both forms, i.e. $A \wedge B$ and $B \wedge A$. The respondents were selected especially among students from Faculty of Information Technology and in total we received 204 questionnaires. This means more than 1000 empirical data. Examples of used statements:

1. The tickets are significantly more expensive.
2. Traveling by public transport is comfortable.
3. The tickets are significantly more expensive, but traveling by public transport is comfortable.
4. Traveling by public transport is comfortable, but the tickets are significantly more expensive.

4   Experimental Results

The main objective of the experiment is a modelling of the conjunction in human language. The second objective is to determine if the human conjunction is a commutative operation. In this paper we consider modelling of the conjunction via t-norms, means and Choquet integral. Collected data show, that respondents understood conjunction as commutative operation only in 56 % of cases. But this result probably depends on the pattern of respondents.

First we focus on modelling via t-norms. Before the approximation collected empirical data are transformed into the interval $[0, 1]$. Since all values from the set $\{1, \ldots, 10\}$ occur in empirical data, the boundary condition ε is set to zero.

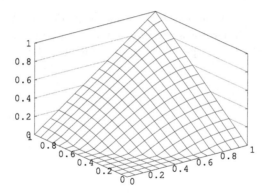

Fig. 2. Graph of a triangular norm that corresponds to conjunction in human use.

Using the algorithm described before, the quadratic additive generator g on the interval $[0,1]$ is given by $g(x) = (1-x)^2$. This generator corresponds to a t-norm

$$T(x,y) = \max\left\{0, 1 - \left((1-x)^2 + (1-y)^2\right)^{\frac{1}{2}}\right\},$$

which is Yager's t-norm with parameter $p = 2$. This t-norm is shown on Fig. 2.

The second considered aggregation operator for modelling conjunction are means. Modified algorithm from Sect. 2.4 is used for mean's generator approximation. Approximated generator g based on measured data is given by $g(x) = x^2$ (Fig. 3b). For approximation, B-spline of degree 3 was chosen. The mean generated by g is in Fig. 3a.

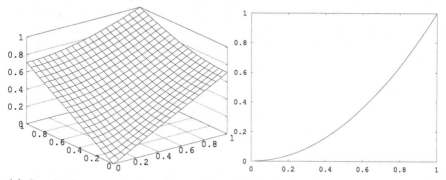

(a) Graph of a mean that corresponds to conjunction in human use.

(b) Generator of a mean approximated by collected data.

Fig. 3. Result of the conjunction modelling via generated mean.

The last mentioned algorithm's modification relates to Choquet based operators. For the generator approximation, B-spline of degree 3 was chosen, as in previous cases. The resulting generator is shown in Fig. 4b. As one can see on Fig. 4a, approximated operator is not a commutative operation. In this way, the approximated operator corresponds mostly to empirical data.

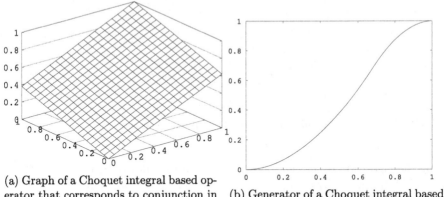

(a) Graph of a Choquet integral based operator that corresponds to conjunction in human use.

(b) Generator of a Choquet integral based operator approximated by collected data.

Fig. 4. Result of the conjunction modelling via Choquet integral.

5 Conclusion

We have tried to model human understanding conjunctions. We tested three different approaches from those trials best reflects reality modelling by Choquet integral. What is important is that we were able to approximate the empirically measured data by aggregation operators. It can be used in many applications and we plan to use our experience with approximation in the recommender systems.

References

1. Hersh, H.M., Caramazza, A.: A fuzzy set approach to modifiers and vagueness in natural language. J. Exp. Psychol. Gen. **105**, 254–276 (1976)
2. Klement, E.P., Mesiar, R., Pap, E.: Triangular Norms. Kluwer Academic Publishers, Boston (2000)
3. Jenei, S., Fodor, J.C.: On continuous triangular norms. Fuzzy Sets Syst. **100**(1–3), 273–282 (1998)
4. Jenei, S.: On archimedean triangular norms. Fuzzy Sets Syst. **99**(2), 179–186 (1998)
5. Beliakov, G., Mesiar, R., Valaskova, L.: Fitting generated aggregation operators to empirical data. Int. J. Uncertainty, Fuzziness Knowl. Based Syst. **12**(2), 219–236 (2004)
6. Beliakov, G.: Fitting triangular norms to empirical data. In: Klement, E., Mesiar, R. (eds.) Analytic and Probabilistic Aspects of Triangular, pp. 261–272. Elsevier, Boston (2005)

7. De Boor, C.: A Practical Guide to Splines. Applied Mathematical Sciences. Springer, New York (2001)
8. Björck, A.: Numerical Methods for Least Squares Problems. Society for Industrial and Applied Mathematics, Philadelphia (1996)
9. Chen, D., Plemmons, R.J.: Nonnegativity constraints in numerical analysis. In: The Birth of Numerical Analysis, pp. 109–139. World Scientific, Singapore (2010)
10. Havlena, V.: Konjunkce a disjunkce ve fuzzy logice. Bakalářská práce, FIT VUT v Brně (2015)
11. Knuth, D.E.: The Art of Computer Programming. Seminumerical Algorithms, vol. 2, 3rd edn. Addison-Wesley Longman Publishing Co., Inc., Boston (1997)
12. Bunch, J.R., Hopcroft, J.E.: Triangular factorization and inversion by fast matrix multiplication. Math. Comput. **28**(125), 231–236 (1974)
13. Žára, J., Beneš, B., Sochor, J.: Moderní počítačová grafika. Computer Press, a.s. (2004)

Practical Exhaustive Generation of Small Multiway Cuts in Sparse Graphs

Petr Hliněný$^{(\boxtimes)}$ and Ondřej Slámečka

Faculty of Informatics, Masaryk University, Botanická 68a,
602 00 Brno, Czech Republic
{hlineny,xslameck}@fi.muni.cz

Abstract. We propose a new algorithm for practically feasible exhaustive generation of small multiway cuts in sparse graphs. The purpose of the algorithm is to support a complete analysis of critical combinations of road disruptions in real-world road networks. Our algorithm elaborates on a simple underlying idea from matroid theory – that a circuit-cocircuit intersection cannot have cardinality one (here cocircuits are the generated cuts). We evaluate the practical performance of the algorithm on real-world road networks, and propose algorithmic improvements based on the technique of generation by a canonical construction path.

1 Introduction

In the area of real-world road network planning and management, one of the vital tasks is to identify potential vulnerabilities of the network in advance. One of the most critical such vulnerabilities is the possibility of a complete break-up of the network as a result of simultaneous disruptions of several roads. In graph theory terms, this corresponds to finding minimal cuts in the network graph (here we consider *edge cuts* by default). However, not every graph cut corresponds to a major disintegration of the whole network; e.g., a cut may just isolate one or several unimportant road intersections (or small villages) and the rest of the network remains fully functional. In fact, one can easily imagine that most small cuts in a real-world network are of the latter (unimportant) kind.

There exist various rather complicated measures of severity of a network break-up, taking into an account the numbers of inhabitants and the economic importance of the nodes which get disconnected from each other, as well as the number of components (cells) into which the network is broken up. See [1] for further references. Notice that, in particular, we have to consider also cuts which separate the network into more than two components (called *multiway cuts*). In a nutshell, research shows that efficient identification of all severe network break-ups does not seem possible without first exhaustively generating all the minimal multiway cuts with small number of edges in the given network.

In our paper we focus right on this task. If we fix integers k, m, then the *task of generating* all the minimal k-way cuts consisting of at most m edges is, in theory, solvable in polynomial time by brute force testing all combinations of at most m

Research supported by the Czech Science Foundation, project 14-03501S.

J. Kofroň and T. Vojnar (Eds.): MEMICS 2015, LNCS 9548, pp. 54–66, 2016.
DOI: 10.1007/978-3-319-29817-7_6

edges of the network. However, experiments carried out over networks of around 1000 edges in [1] clearly show that such a brute force approach is practically feasible, even on parallel machines, only for $m \leq 4$. To support the analysis of road network break-ups caused by more than 4 simultaneous disruptions, we proposed a new approach to an exhaustive small cut generation whose simplified heuristic version has already been implemented and successfully used in [1].

The underlying idea of our proposed approach is very natural and simple: suppose we construct a (to-be) cut X iteratively, and there exists a cycle C such that C intersects X in exactly one edge—then another edge of C must belong to the resulting minimal cut extending X. Consequently, the next iteration can choose only from the edges of C instead of the whole network. Since in real-world road networks one can usually find an abundance of short cycles everywhere (where "short" typically means 4 or 5), this approach can dramatically reduce the search space and the runtime of the algorithm.

Here we provide a theoretical background for this new *Circuit-cocircuit algorithm scheme* in terms of matroid theory, which seamlessly integrates generation of minimal k-way cuts for all values of k into the one scheme. We further elaborate the algorithm towards the so-called canonical generation which provides additional important speed-up. We also report on the results of practical computational experiments carried out with different versions of our algorithm.

Related research. Computing a *minimum* two-terminal cut in a graph is a well-known easy application of network flow theory. However, the seemingly similar problem of counting the minimum cuts in a graph is #P-complete [7] (i.e., equivalent to #SAT). Notice that there is a crucial difference between the terms *minimum* and *minimal* cut—where "minimum" means of smallest possible cardinality and "minimal" cuts are those for which no proper subset of them is a cut again (while their cardinality may be arbitrarily high). In our task, if we set m equal to the minimum cut size in the graph (instead of fixing it to a small value beforehand), we hence get that our generation problem is #P-hard in general. Fortunately, experiments with real-world road networks show that their particular case is often computationally much simpler.

Concerning k-terminal cuts for $k > 2$, already computing a minimum three-terminal cut in a graph is an APX-hard problem [3]. Consequently, things do not get any easier with generating multiway cuts. Besides obvious brute force attempts, not much has been published in literature about exhaustive generation of small cuts in graphs. One remarkable exception is the work of Reinelt and Wenger [8], who elaborated on the classical so-called "cactus representation" of Dinitz et al. [4] to provide a practically efficient algorithm for generation of all minimum multiway cuts in a graph. However, the problem with [8] and previous related papers is that they all compute "minimum" cuts, but in our case we have to generate also all the larger minimal cuts (up to a cardinality bound m) in addition to the minimum (in terms of cardinality) ones.

Paper organization. In Sect. 2 we give a brief introduction to the necessary theoretical concepts. After that we state the abstract Circuit-cocircuit algorithm for matroids (Algorithm 3) and illustrate a simple use of it for generating minimal 2-cuts in a graph (Algorithm 5). The full power of the Circuit-cocircuit meta-algorithm shows up in Sect. 4 where we apply it to exhaustively generate all minimal k-way cuts in a graph (Algorithm 7). Section 5 then outlines a further improvement of the algorithm using the so-called canonical generation.

2   Preliminaries

We mostly follow standard terminology of graph theory. The vertex set of a graph G is referred to as $V(G)$ and the edge set as $E(G)$. In the paper we pay a particular attention to the following graph terms.

An *edge-cut* in a graph G is a set of edges $X \subseteq E(G)$ such that $G \setminus X$ (the subgraph of G obtained by deleting the edges X) has more connected components than G has. A *k-way edge-cut* in a graph G is a set of edges $X \subseteq E(G)$ such that $G \setminus X$ has at least k connected components. Note that in connected graphs, an edge-cut coincides with a 2-way edge-cut, while in a disconnected graph this assertion fails (the empty set is then a 2-way edge-cut).

Definition 1 (Bond). *We call a* bond *any minimal edge-cut in a graph, and a* k-bond *any minimal k-way edge-cut in a graph (minimality is considered with respect to set inclusion).*

A graph is a *tree* if it is connected but deleting any of its edges disconnects it. In other words, a tree contains no cycles. A graph is a *forest* if each of its connected components is a tree. If G is a graph and $F \subseteq G$ is a tree (forest) such that $V(F) = V(G)$, then F is a *spanning tree (forest)* of G.

It turns out that the most suitable framework for an abstract description of our proposed algorithm is that of matroid theory. We follow Oxley [6] in matroid terminology, and we give a brief introduction (with examples) next.

Definition 2 (Matroid). *A matroid is a pair $M = (E, \mathcal{B})$ where $E = E(M)$ is the finite ground set of M (elements of M), and $\mathcal{B} \subseteq 2^E$ is a nonempty collection of* bases *of M, no two of which are in an inclusion. Moreover, matroid bases must satisfy the "exchange axiom"; if $B_1, B_2 \in \mathcal{B}$ and $x \in B_1 \setminus B_2$, then there is $y \in B_2 \setminus B_1$ such that $(B_1 \setminus \{x\}) \cup \{y\} \in \mathcal{B}$.*

The following terminology is used in matroid theory. Subsets of bases are called *independent sets*, and the remaining sets are *dependent*. Minimal sets not contained in a basis (i.e., dependent sets) are called *circuits*, and maximal sets not containing any basis are called *hyperplanes*.

Example 1. Let $A = \{a_1, \ldots, a_n\}$ be a finite set of vectors. If $\mathcal{B}$ is the set of all maximal independent subsets of A, then $M = (A, \mathcal{B})$ is a matroid, called the *vector matroid* of A. The independent sets of M are precisely the linearly independent subsets of A.

Example 2. If G is a connected graph, then its *cycle matroid* on the ground set $E(G)$ is as follows: bases are the (edge sets of the) spanning trees of G, independent sets are the (edge sets of) forests in G, circuits are the usual cycles in G, and hyperplanes are the set complements of bonds in G.

For a matroid $M = (E, \mathcal{B})$, the matroid on the same ground set E and with the (complementary) bases $B^* = \{E \setminus B : B \in \mathcal{B}\}$ is called the *dual matroid of* M and denoted by M^*. The circuits of M^* are called *cocircuits of M*.

Example 3. Let G be a planar graph and M the cycle matroid of G. Then M^* is the cycle matroid of the geometric dual of G.

Claim 1 (folklore, see [6]). *Let M be a matroid.*

(a) If B is a basis of M and $e \in E \setminus B$, then $B \cup \{e\}$ contains precisely one circuit (through e).

(b) If H is a hyperplane of M and $e \in E \setminus H$, then $H \cup \{e\}$ contains a basis B of M and $e \in B$.

(c) A set $X \subseteq E$ is a cocircuit of M iff $E \setminus X$ is a hyperplane of M.

(d) Cocircuits of M are precisely the minimal sets intersecting every basis of M.

Claim 2 (cf. Example 2). *Let M be the cycle matroid of a graph G. Then the cocircuits of M are precisely the bonds of G.* □

3 The Circuit-Cocircuit Meta-Algorithm

In view of Claim 2, it is possible to formulate the problem of generating all bonds of a graph as generating all the cocircuits of its cycle matroid. This approach might seem restrictive at the first sight as it does not directly capture generation of k-bonds for $k > 2$, but precisely the opposite is true: we will later show that k-bonds are the cocircuits under a suitably adjusted definition of the cycle matroid of a graph.

It is quite natural to see that a cycle and a bond in a graph cannot intersect in precisely one edge. A generalization of this observation is one of the fundamental claims in matroid theory (note, however, that a matroid circuit and a cocircuit may intersect in 3 or 5, etc., elements...):

Proposition 1 (folklore, see [6]). *If C is a circuit and X is a cocircuit in a matroid M, then $|C \cap X| \neq 1$.*

With Proposition 1 at hand, we may simply proceed as follows: start with any element of M in X, find a circuit C such that $|C \cap X| = 1$, and then for each element $c \in C \setminus X$ try to add c to X and recurse. The recursion proceeds as long as $E \setminus X$ contains a hyperplane of M, cf. Claim 1(c). The full pseudocode is given in Algorithm 3.

Theorem 4. *Algorithm 3 generates all the cocircuits of size $\leq m$ in a matroid M (with repetition – the same cocircuit may be generated several times).*

Algorithm 3. Abstract Circuit-Cocircuit Meta-algorithm

Input: Matroid $M = (E, \mathcal{B})$ and an integer $m \in \mathbb{N}$ (a cocircuit size bound)
Output: All cocircuits of M with size $\leq m$
1:   $B \leftarrow$ an arbitrary basis in $\mathcal{B}$
2: **for all** $b \in B$ **do**
3:     $X \leftarrow \{b\}$
4:     GENCOCIRCUITS(X)
5: **end for**
6: **procedure** GENCOCIRCUITS(X)
7:     **if** $E \setminus X$ contains no hyperplane of M **or** $|X| > m$ **then**
8:       **return** $\perp$ ▷ this branch fails
9:     **end if**
10:    Find any circuit $C \subseteq E$ such that $|C \cap X| = 1$
11:    **if** such C doesn't exist **then**
12:      **output** X ▷ X is a cocircuit
13:    **else**
14:      $D \leftarrow C \setminus X$
15:      **for all** $c \in D$ **do**
16:        GENCOCIRCUITS($X \cup \{c\}$)
17:      **end for**
18:    **end if**
19: **end procedure**

The proof follows rather straightforwardly (though not shortly) from Claim 1, but due to space restrictions it is skipped here.

Remark 1. Note that Algorithm 3 makes some nondeterministic steps – the choices (of B, C) on lines 1,10 and also the ordering (of D) on line 15. Theorem 4 asserts that for any particular implementation of these steps, the algorithm remains correct. We exploit this fact mainly with the choice of C on line 10, where we aim to minimize $|C|$. If we are (mostly) able to choose C "very small", bounded by a constant such as 5 or 6, then we get a dramatic runtime speed-up over the basic brute force approach trying all $\leq m$-elements subsets of E. Indeed, this is the typical case for the cycle matroids of real-world road networks.

Remark 2. There is one weakness of Algorithm 3 which is common to many iterative/recursive combinatorial generation algorithms—the same object (here a cocircuit or a bond) is generated many times in different orders of its elements. While there is no easy general remedy for this common problem, we will provide a practically working fast resolution in Sect. 5.

3.1 Generating 2-Bonds in a Graph

To better explain Algorithm 3 and its use, we now present a sample implementation for generating all the 2-bonds in a connected graph. The main task of our implementation is to realize line 7—to be able to efficiently test whether $E \setminus X$ contains a hyperplane of M. This is based on the following claim:

Lemma 1. *Let G be a connected graph, $M = (E, \mathcal{B})$ its cycle matroid and $Y \subseteq E = E(G)$. The set $E \setminus Y$ contains a hyperplane of M if, and only if, the vertices incident to the edges of Y can be coloured red and blue, such that each edge of Y gets two colours and there exist two disjoint trees T_r and T_b in $G \setminus Y$ such that the tree T_r (T_b) connects all the red (blue, resp.) vertices of Y.*

Proof. (=>) If $E \setminus Y$ contains a hyperplane of M, then there exists a cocircuit $X \supseteq Y$ by Claim 1(c). Since X is a 2-bond in G, $G \setminus X$ has precisely two connected components (as otherwise X would not be minimal). Colouring the ends of Y in one component red and in the other blue finishes the argument.

(<=) Let $R = V(T_r)$ and $B = V(T_b)$ be the vertex sets of the assumed two trees in $G \setminus Y$. Let $U \subseteq V(G)$ be the set reachable from R in $G \setminus B$ and $X \subseteq E(G)$ be the edges having precisely one end in U. Then $Y \subseteq X$ by the definition, and X is a cut in G separating R from B. Moreover, X is minimal, and so X is a 2-bond and $E \setminus X$ is a hyperplane of M which is contained in $E \setminus Y$.

□

In regard of Lemma 1, we choose the following implementation of the hyperplane test on line 7. During the progress of the algorithm, each edge e chosen to be added to X gets the colours *red* and *blue* at its ends, such that this choice is consistent (wrt. edges already in X) and fulfills the next conditions (Algorithm 5). This implementation results in the following algorithm:

Algorithm 5 (Circuit-Cocircuit algorithm for 2-bonds in a graph).
We specify Algorithm 1 with the following points:

(1) Let M of Algorithm 1 be the cycle matroid of an input graph G.
(2) With respect to implementation of line 7, the first edge added to X on line 3 gets the colours red/blue arbitrarily. Let, subsequently, $V(X) = V_r \cup V_b$ where V_r are the red ends of X and V_b the blue ends. We actively maintain only a *red tree* T_r interconnecting V_r (as expected by Lemma 1), while a *blue tree* is implicit – the two trees are not treated symmetrically: see further Algorithm 3 for details of building and maintaining T_r.
(3) Instead of a cycle C on line 10, we explicitly look for a (shortest) path $P \subseteq G \setminus X$ such that one end of P is $u_r \in V_r$ and the other end is $u_b \in V_b$. Note that, for any $f \in X$ with one end u_b, there is a cycle C formed by P, f and the unique path in T_r from u_r to f such that $C \cap X = \{f\}$, as expected by Algorithm 1, but we do not explicitly invoke C in our implementation.
(4) On line 14, we set $D \leftarrow E(P)$ (which is a subset of the implicit circuit C).

Proposition 2. *Algorithm 5 generates all the 2-bonds of size $\leq m$ in a connected graph G (with possible repetition).*

The proof nearly immediately follows from Theorem 4 and Claim 2, but there is one catch: the set D computed on line 14 may be a strict subset of $C \setminus X$ expected in Algorithm 3. We can show that for every 2-bond X_0 of G, at least one of the computation paths leading to X_0 is not affected by this deficiency. Again, due to space restrictions a full proof is skipped here.

4 k-Way Cycle Matroid and Generating k-Bonds

As mentioned before, Algorithm 3 can be used for generating k-bonds of a graph for any $k \geq 2$. We just have to extend the definition of a cycle matroid so that cocircuits within the new definition are precisely the k-bonds.

Definition 3 (k-way cycle matroid). *Let G be a graph of less than $k \geq 2$ components. The k-way cycle matroid of G is a matroid on the ground set $E(G)$, such that its bases are the edge sets of the spanning forests of G consisting of $k-1$ trees. The bases, circuits, cocircuits, hyperplanes of the k-way cycle matroid are also called the k-way bases, circuits, cocircuits, hyperplanes of G.*

From this definition one can easily conclude some basic properties.

Claim 6. *Let G be a graph consisting of less than $k \geq 2$ components. The k-way cocircuits of G are precisely the k-bonds of G. The k-way circuits of G are of two types,* type-C *and* type-F:

– *type-C circuits are the graph cycles in G.*
– *type-F circuits, also called* spanning circuits, *for $k \geq 3$, are the spanning forests of G that are formed by $k-2$ trees.*

Now, by Theorem 4, every implementation of Algorithm 3 for the k-way cycle matroid of a graph G generates all the k-bonds of G. Although, working with the circuits of Claim 6 is somehow intricate. We thus restrict our attention to a special variant of Algorithm 3 which has several advantages.

– First, this variant is compatible with and extends Algorithm 5.
– Second, it coincides with the natural naive approach to generating k-bonds: find a 2-cut, choose one of its sides and recursively find a 2-cut of this side, and so on until k parts are generated. In other words, we also prove that such a naive approach is indeed correct (if properly implemented).

This special variant is defined as follows:

Definition 4 (Stepwise Circuit-Cocircuit implementation scheme). *We call an implementation of Algorithm 3 stepwise if, for every set $X = X_0$, $|X_0| = l$, generated by the algorithm the following holds:*

1. *X_0 is an ordered sequence $(c_1, c_2, \ldots, c_l)$, where c_i has been added to X_0 at the level $i-1$ of recursion, and*
2. *there exists a mapping $s : \{1, 2, \ldots, k\} \to \{0, 1, \ldots, l\}$ such that $s(1) = 0$, $s(k) = l$ and, for each $j \in \{2, \ldots, k-1\}$, the set $\{c_1, \ldots, c_{s(j)}\} \subsetneq X_0$ forms a j-bond in G.*

For j, $1 \leq j < k$, we call the j-th stage of the algorithm the steps the algorithm does at the levels $s(j), s(j)+1, \ldots, s(j+1)-1$ of recursion. In other words, the algorithm in its j-th stage selects the elements $c_{s(j)+1}, \ldots, c_{s(j+1)}$.

Before proceeding into details of the stepwise implementations, we first show that the definition indeed makes sense.

Algorithm 7. One stage of a stepwise implementation

Input: A conn. graph G, param. $j, k, m \in \mathbb{N}, j < k, m \geq 1$, and a j-bond $Y_1 \subseteq E(G)$
Output: A collection of $(j+1)$-bonds such that for each k-bond Y, $Y_1 \subseteq Y \subseteq E(G)$, $|Y| \leq m$, some subset of Y is among the generated $(j+1)$-bonds

1: **if** $j = 1$ **then** $\qquad\qquad\qquad\qquad\qquad$ ▷ $Y_1 = \emptyset$: select a k-way basis
2: $\qquad$ $F \leftarrow$ an arb. spanning forest of $k - 1$ trees
3: **else** $\qquad\qquad\qquad\qquad\qquad\qquad\qquad$ ▷ $Y_1 \neq \emptyset$: select a type-F circuit
4: $\qquad$ $F \leftarrow$ an arb. spanning forest of $k - 2$ trees and $|F \cap Y_1| = 1$
5: **end if**
6: **for all** $d = \{u, v\} \in F \setminus Y_1$ **do**
7: $\qquad$ GENSTAGE$(j, Y_1, X = \{d\}, V_r = \{u\}, V_b = \{v\}, T_r = \{u\})$
8: **end for.**

9: **procedure** GENSTAGE(j, Y, X, V_r, V_b, T_r)
10: $\qquad$ Let $G_1 \subseteq G$ be the component of $G \setminus Y$ containing X
11: $\qquad$ **if** $|Y \cup X| > m - k + j + 1$ **then**
12: $\qquad\qquad$ **return** $\bot$ $\qquad\qquad\qquad$ ▷ no way to get a k-bond of size $\leq m$
13: $\qquad$ **end if**
14: $\qquad$ **if** there does not exist a connected subgraph
15: $\qquad\qquad$ $T_b \subseteq (G_1 \setminus V(T_r)) \setminus X$ such that $V_b \subsetneq V(T_b)$ **then**
16: $\qquad\qquad$ **return** $\bot$ $\qquad\qquad\qquad$ ▷ the "no hyperplane" condition
17: $\qquad$ **end if**
18: $\qquad$ $P \leftarrow$ a minimal path in G_1 from $V(T_r)$ to V_b
19: $\qquad$ **if** such P does not exist **then**
20: $\qquad\qquad$ **output** $Y \cup X$ $\qquad\qquad\qquad$ ▷ $Y \cup X$ is a $j + 1$-bond
21: $\qquad$ **else**
22: $\qquad\qquad$ **for all** $c \in P$ **do** $\qquad\qquad$ ▷ add c to X and update T_r
23: $\qquad\qquad\qquad$ Let u be the vertex in $c = \{u, v\}$ which is closer to T_r
24: $\qquad\qquad\qquad$ Let P_u be the component of $P - c$ which contains u
25: $\qquad\qquad\qquad$ GENSTAGE$(j, Y, X \cup \{c\}, V_r \cup \{u\}, V_b \cup \{v\}, T_r \cup P_u)$
26: $\qquad\qquad$ **end for**
27: $\qquad$ **end if**
28: **end procedure**

Proposition 3. *A stepwise implementation of Algorithm 3 is possible. Precisely, for every $k \geq 2$ there exists a stepwise implementation generating all the k-bonds in a given connected graph.*

One can, moreover, easily show that a "transition" from the j-th stage to $(j+1)$-st one in a stepwise implementation really means to construct a 2-bond in one of the parts of the previous j-bond. A desired consequence is that we can decompose the stepwise algorithm computation into these stages such that, in each stage, we simply invoke Algorithm 5.

These findings directly lead to a stepwise algorithm whose one stage is shown in pseudocode in Algorithm 7. Validity of this new algorithm then, in turn, follows immediately from the following statement describing its one stage output.

Theorem 8. *Let G be a graph, j, k, m integers such that $j < k, m \geq 1$ and $Y_1 \subseteq E(G)$ a j-bond in G. Algorithm 7 generates a set $\mathcal{S}$ of $(j+1)$-bonds such* *that for each k-bond Y, $Y_1 \subseteq Y \subseteq E(G)$, $|Y| \leq m$, some subset of Y is among the generated $(j + 1)$-bonds in $\mathcal{S}$.*

The proof follows from Proposition 2 via the previous claims.

5   Canonical Generation

We now return to Remark 2; addressing the problem that one bond X_0 is typically generated many times by our circuit-cocircuit algorithm, each time with a different permutation of its elements. While such a repetition can be easily removed by a post-processing, it costs running time. Ideally, our algorithm should for each X_0 "guess" one computation path leading to X_0 and immediately dismiss all the other attempts, as early as possible in the generation process. But how can this be done? This is not at all an easy question since, for example, we have to ensure that (nearly) every two bonds X_0, X_1 sharing many elements also share a long prefix of the guessed computation path, and so on. Most importantly, the guessed computation path of X_0 must be compatible with Algorithm 3, i.e., each next element of X_0 on the path must be from the circuit C on line 10 of the algorithm, which is not a priori clear how to achieve.

There exists a sophisticated technique of *generation by a canonical construction path* by McKay [5], outlined next. Since we cannot fit the details of this technique and its application to our case into the restricted conference paper, we stay on a very informal level.

In our case, a computation path of a bond X_0 in G is simply encoded by a permutation $\vec{X}_0$ of the elements of X_0. The definition of a canonical form $\vec{X}_0$ of X_0 respects the stepwise generation framework as follows:

(I)   The permutation $\vec{X}_0$ refines the order of the stages in some stepwise computation path leading to X_0 (cf. Definition 4).

(II) There is an arbitrary bijection $\iota : E(G) \rightarrow \{1, \ldots, |E(G)|\}$ indexing the edges of G. The starting edges of the stages in $\vec{X}_0$ are each ι-minimal within its stage, and they are altogether strictly ordered by ι (first-to-last stage).

(III) Within each stage, the corresponding sub-permutation of $\vec{X}_0$ (except the starting edge) is determined by the shortest path P selection and the red/blue tree mechanism of Algorithm 5; see also the appropriate parts of Algorithm 7.

Two additional details are important for a successful implementation of this point. First, the red and blue sides of the hyperplane test are uniquely decided with the first edge of the stage based on a fixed vertex indexing of G. Second, the unit lengths of edges of G are slightly perturbed to achieve uniqueness of the shortest path P selection.

Concerning the canonical implementation of bond generation, point (I) and parts of (III) of the scheme are already embedded in Algorithm 7, and the rest of (III) is rather straightforward to add. The biggest runtime savings come from implementing point (II). At the beginning of each stage, the starting edge is selected from an ι-minimal basis (or type-F circuit) among its edges of ι-value higher than that of the previous stage. Then, the remaining edges of this stage are restricted only to those candidates of higher ι-value than the starting edge.

Although the presented scheme is not *truly canonical* since one k-bond X_0 can still be generated in more than one canonical form, it is implementation-wise very easy and provides great speed-up for the algorithm; see the next section.

6 Evaluation

In this section we present the outcomes of measurements performed with implementations of our algorithms on the road networks of the regions of Czech republic: the Zlín Region (723 vertices, 974 edges) and the Olomouc Region (1454 vertices, 2066 edges). Measurements using the larger road network of Central Bohemian Region (4114 vertices, 5964 edges) gave similar results.

We have implemented the core algorithm of Sect. 4 which generates same bonds multiple times (i.e., without canonical generation), and the improved algorithm of canonical generation from Sect. 5. For the running time evaluation we used a computer with 16 GB RAM and the Intel Core i7-3770 CPU @ 3.40 GHz. The source code was compiled with gcc 4.8.2.

The measurement results are summarized in the tables below. To start, Tables 1 and 2 show the overall runtimes where the entries marked '-' did not finish before the time limit. Tables 3 and 4 show the improvement, in terms of runtime, of the canonical generation algorithm from Sect. 5 over the ordinary algorithm from Sect. 4. The improvement achieved by preventing repeated generation of the same bonds is up to 15× in the experiments. This runtime improvement well correlates with the average multiplicity of repeatedly generated bonds by the ordinary algorithm in Table 5. Although the approach of Sect. 5 does not completely prevent repeated generation of the same bonds, the percentage of "leftover" multiply generated bonds is truly marginal and hence negligible for practical computations; see Table 6.

To demonstrate superiority of the circuit-cocircuit algorithm over the brute-force approach trying all m-tuples of edges for k-bonds, we include Table 7. The table summarizes the distribution of lengths of the path P (Algorithm 7, line 18), which represent the degrees of branching of the circuit-cocircuit algorithm inside each stage. While the brute-force approach would result in a quite bad running time of order $\mathcal{O}(|E(G)|^m/m!)$, the nature of Algorithm 7 together with the experimental data in Table 7 suggest that the running time can be, roughly,

$$\mathcal{O}\big(|V(G)|^k \cdot \beta^{m-k}\big), \tag{1}$$

where the auxiliary constant β stands for a typical bound on the length of the path P and can be guessed as $\beta \approx 5$.

Comparing to Tables 1 and 2, one can see quite a good match in the runtime dependence on $(m-k)$ in (1), while the dependence on k seems overshadowed by other aspects of the algorithm for the small experimental values of k, m.

Table 1. Running time of an implementation of the canonical generation in seconds. Zlín Region

m / k	2	3	4	5	6	7	8
2	0.0	0.1	1	2	9	42	210
3	0.6	2.8	13	53	223	986	4604
4		29.5	198	1018	4771	21269	-
5			1156	9885	56847	-	-

Table 2. Running time of an implementation of the canonical generation in seconds. Olomouc Region

m / k	2	3	4	5	6	7	8
2	0.1	0.3	1	5	16	69	305
3	3.0	10.4	61	235	921	3482	13342
4		158.3	781	6008	-	-	-
5			6205	43242	-	-	-

Table 3. Ratio of running times without and with canonical generation. Zlín (top) and Olomouc (bottom) Region

m / k	2	3	4	5	6	7
2	1.00	2.00	2.45	3.60	4.3	1.34
3	3.28	3.51	6.05	8.83	12.11	14.90
4		8.40	11.32	15.73	-	-

m / k	2	3	4	5	6	7
2	1.14	1.93	1.91	2.18	4.30	5.06
3	1.97	3.47	4.49	6.69	8.31	9.41
4		5.96	10.16	15.53	-	-

Table 4. Ratio of the numbers of generated bonds without and with canonical generation. Zlín (top) and Olomouc (bottom) Region

m / k	2	3	4	5	6	7
2	1.43	1.97	2.67	3.52	4.42	1.15
3	2.00	3.08	4.27	5.99	7.94	10.09
4		6.00	9.50	13.38	-	-

m / k	2	3	4	5	6	7
2	1.32	1.93	2.53	3.26	4.04	4.78
3	2.00	2.84	4.03	5.50	7.41	9.59
4		6.00	8.79	12.37	-	-

Table 5. The average multiplicity of (repeatedly) generated bonds in the non-canonical generation algorithm. Zlín (top) and Olomouc (bottom)

m / k	3	4	5	6	7
3	3.112	4.309	6.092	8.152	10.465
4		9.706	13.642	-	-

m / k	3	4	5	6	7
3	2.840	4.038	5.536	7.491	9.739
4		8.817	12.432	-	-

Table 6. The percentage of repeatedly generated bonds in the canonical generation algorithm. Zlín (top) and Olomouc (bottom) Region

m / k	3	4	5	6	7
3	0.972%	0.814%	1.618%	2.664%	3.715%
4		2.177%	1.950%	3.462%	-

m / k	3	4	5	6	7
3	0.156%	0.253%	0.649%	1.049%	1.568%
4		0.352%	0.541%	-	-

Lastly, we would like to comment on a possible *parallelization* of the new algorithm. This is actually very easy: each time when adding a new edge to the bond X, one may simply run all the computation branches in parallel, without *any need* for synchronization or communication between the branches. Furthermore, especially

Table 7. The distribution of lengths of the path P from Algorithm 7. Results of the computation on Zlín Region, $k = 3, m = 6$; on the left showing the second level of recursion of GenStage, on the right the fifth level (the algorithm occasionally uses even longer paths in later GenStage calls).

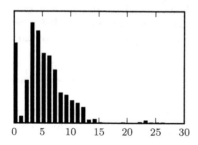 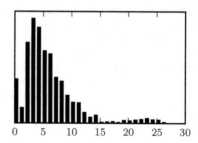

in the canonical generation case, no costly final post-processing of the generated bonds is needed.

7   Conclusion

We have presented a new "Circuit-Cocircuit" algorithm for exhaustive generation of cocircuits in a matroid, with a practical application to finding all the minimal k-way cuts in a graph. We have further elaborated on the algorithm to achieve an almost canonical generation process, which significantly speeds-up the algorithm by early removal of duplicate computation branches. This theoretical work has been complemented by an implementation and extensive practical evaluations of the algorithm on real-world data. The source code of our implementation is available at https://github.com/OndrejSlamecka/mincuts.

In a conclusion, our implementation solves the problem of finding all small multiway cuts correctly as well as quickly (given the high theoretical complexity of the problem) and with very low memory usage, thus demonstrating the feasibility of this algorithm for practical computations, e.g., in road network planning and management. In particular, the algorithm performs significantly better than the brute-force algorithm on real-world networks. Our algorithm will help to improve the results of [1] (where only a simplified heuristic version of the Circuit-Cocircuit algorithm, without canonicity, was implemented).

Our main suggestions for future work are as follows. The main theoretical question is whether there exists a method of truly canonical generation which does not require costly explicit isomorphism checks. On the implementation side, profiling shows that the algorithm spends most of time in the shortestPath procedure—finding a good CPU-aware implementation [2] of this procedure would benefit the running time.

References

1. Bíl, M., Vodák, R., Hliněný, P., Svoboda, T., Rebok, T.: A novel method for rapid identification of road links causing network break-up, submitted to EJOR (2015)
2. Chhugani, J., Satish, N., Kim, C., Sewall, J., Dubey, P.: Fast and efficient graph traversal algorithm for CPUs: Maximizing single-node efficiency. In: 2012 IEEE 26th International Parallel and Distributed Processing Symposium. IEEE, May 2012. http://dx.org/10.1109/IPDPS.2012.43
3. Dahlhaus, E., Johnson, D.S., Papadimitriou, C.H., Seymour, P.D., Yannakakis, M.: The complexity of multiterminal cuts. SIAM J. Comput. 23(4), 864–894 (1994). http://dx.doi.org/10.1137/S0097539792225297
4. Dinic, E., Karzanov, A., Lomonosov, M.: A structure of the system of all minimum cuts of a graph. In: Fridman, A.A. (ed.) Studies in Discrete Optimization, Nauka, Moscow, pp. 290–306 (in Russian) (1976)
5. McKay, B.D.: Isomorph-free exhaustive generation. J. Algorithms 26(2), 306–324 (1998). http://dx.org/10.1006/jagm.1997.0898
6. Oxley, J.: Matroid Theory. Oxford graduate texts in mathematics. Oxford University Press, New York, USA (2006). http://books.google.cz/books?id=puKta1Hdz-8C
7. Provan, J.S., Ball, M.O.: The complexity of counting cuts and of computing the probability that a graph is connected. SIAM J. Comput. 12(4), 777–788 (1983). http://dx.org/10.1137/0212053
8. Reinelt, G., Wenger, K.M.: Generating partitions of a graph into a fixed number of minimum weight cuts. Discrete Optim. 7(1–2), 1–12 (2010). http://dx.org/10.1016/j.disopt.2009.07.001

Self-Adaptive Architecture for Multi-Sensor Embedded Vision System

Ali Isavudeen[1,2]([⊠]), Eva Dokladalova[1], Nicolas Ngan[2], and Mohamed Akil[1]

[1] Laboratoire Informatique Gaspard Monge, Equipe A3SI Unité Mixte
CNRS-UMLV-ESIEE (UMR 8049), Noisy-le-Grand, France
{eva.dokladalova,mohamed.akil}@esiee.fr
[2] Sagem Défense et Sécurité Groupe Safran, Argenteuil, France
{ali.isavudeen,nicolas.ngan}@sagem.com

Abstract. Architectural optimization for heterogeneous multi-sensor processing is a real technological challenge. Most of the vision systems involve only one single color sensor and they do not address the heterogeneous sensors challenge. However, more and more applications require other types of sensor, in addition, such as infrared or low-light sensor, so that the vision system could face various luminosity conditions. These heterogeneous sensors could differ in the spectral band, the resolution or even the frame rate. Such sensor variety needs huge computing performance, but embedded systems have stringent area and power constraints. Reconfigurable architecture makes possible flexible computing while respecting the latter constraints. Many reconfigurable architectures for vision application have been proposed in the past. Yet, few of them propose a real dynamic adaptation capability to manage sensor heterogeneity. In this paper, a self-adaptive architecture is proposed to deal with heterogeneous sensors dynamically. This architecture supports on-the-fly sensor switch. The architecture of the system is self-adapted thanks to a system monitor and an adaptation controller. A stream header concept is used to convey sensor information to the self-adaptive architecture. The proposed architecture was implemented in Altera Cyclone V FPGA. In this implementation, adaptation of the architecture consists in Dynamic and Partial Reconfiguration of FPGA. The self-adaptive ability of the architecture has been proved with low resource overhead and an average global adaptation time of 75 ms.

1 Introduction

The performance capability of modern embedded vision system is increasing day by day. Requirements of a vision system mostly depend on the application of the system, and priorities on performance criteria will be different whether it is for public, automotive, medical or military purpose. In military application, for instance, vision systems have to deal with multiple and various environmental and operational contexts. They should handle all luminosity conditions, be it day, night, indoor or outdoor environment. The operational context could either be surveillance, tracking or targeting.

© Springer International Publishing Switzerland 2016
J. Kofroň and T. Vojnar (Eds.): MEMICS 2015, LNCS 9548, pp. 67–78, 2016.
DOI: 10.1007/978-3-319-29817-7_7

Most of time, both industrial and academic architecture of embedded vision system integrate only one single image sensor. The widely used and known one is the CMOS color sensor. This sensor can capture and produce a color digital picture [1]. However, the CMOS color sensor is not sufficient to face the variable environmental and luminosity context. The vision system has to integrate different types of sensor such as color, infrared or intensified light sensor. These sensors differ in many characteristics such as the spectral band, the Field of View (FOV), the frame resolution, the frame rate or even the pixel size. This sensor variety makes difficult the architecture designing. Actually, in the case of heterogeneous sensors vision system, the processing architecture has to carry out different types of data at different frequencies. While general purpose processor-based systems are often designed for uniform data processing, reconfigurable computing [2] offers the possibility to carry out heterogeneous data processing.

Many reconfigurable hardware-based or reprogrammable software-based solutions have been proposed previously [3–5]. Nevertheless, these architectures do not support dynamic replacement of the sensor. The real challenge that we want to tackle here is the dynamic adaptability of the architecture in a heterogeneous multi-sensor vision system. The expected system should be able to adapt its processing architecture dynamically to manage the on-the-fly replacement of the sensor.

In this paper, we propose a novel self-adaptive architecture for heterogeneous sensors vision system. The architecture can self-adapt its hardware organization in response to a dynamic switch of the sensor. A stream header conveying information about the sensor is included in the image stream. This information is used to adapt the processing architecture according to the sensor characteristics.

The paper is organized as follows. An overview of related works is given in Sect. 2. The new self-adaptive architecture and its features are presented in Sect. 3 while the experimental prototype for evaluation is presented in Sect. 4. Experimental results are discussed in Sect. 5. Finally, Sect. 6 concludes and announces perspectives for future work.

2   Multi-sensor Embedded Vision System

Photo and video camera are the most widely known vision systems. In these systems, as like as in most of the vision systems, either academic or industrial one, there is only one CMOS color sensor. The CMOS sensor is good enough for day vision purpose, but it is not sufficient for variable luminosity condition. Several other sensors, such as infrared or low-light sensor, are used for medical, automotive or military applications. Infrared sensors, also known as thermal sensors, are sensible to the infrared waves which stem from heat objects. The specificity of the low-light sensors is their ability to work in very low luminosity condition.

The multi-sensor concept has also been used in stereo vision systems [6,7]. In these vision systems, there are two image sensors, but both of them have the same characteristics. They also have the same image processing for both sensors and the processing for both sensors is done in parallel. Our aim is to propose an adaptive processing architecture for vision system involving different kind of sensor.

Processing architecture in many embedded vision system is based on a software solution with a DSP or general purpose processor. Most of the embedded vision systems use an ARM-based video processing core, with an additional DSP and specific image&video co-processing cores [8]. These solutions are mostly designed for a given sensor, which is often a color sensor. Their poor flexibility performance is not suitable for multiple heterogeneous sensor vision systems.

Reconfigurable hardware, such as Field Programmable Gate Arrays (FPGA), are good trade-off for high performance, high flexibility and reasonable power consumption. Many works have explored FPGA-based reconfigurable architecture for vision application [3–5]. In [4], Dynamic and Partial Reconfiguration (DPR) is used to evaluate acceleration performance of some image processing in FPGA implementation. In [5], authors explore DPR to implement an automatic white balance algorithm. Both works do not deal with heterogeneous sensors.

The DreamCam proposed in [3] presents an FPGA implementation for Harris&Stephens corner and edge detection algorithm. This work proposes a reconfigurable solution for embedded cameras involving heterogeneous sensors. However, the architecture can be only statically reconfigured when the sensor is switched. For a given image processing chain, the architecture enables only parameter modifications. The sensor can not be switched dynamically.

Our work attempts to provide a self-adaptive architecture that supports vision system involving multiple and heterogeneous sensors. This architecture can dynamically adapt its organization to enable dynamic switch of the sensor.

3   Self-Adaptive Multi-Sensor System

A multi-sensor vision system involves more than one image sensor. There are many types of sensor that a vision system could integrate. Table 1 gives some examples of existing sensors and their characteristics.

Table 1. Example of sensors characteristics

Type	Spectral band	Color space	Resolution	Frame rate (fps)	Pixel size (bit)
Color [9]	visible	RGB	1280×960	45	3×12
Low-light [10]	visible	Grayscale	1280×1024	60	10
Infrared [11]	infrared	Grayscale	640×480	120	12

The proposed self-adaptive architecture is supposed to manage such a variety of sensors mentioned in Table 1. In our framework, we suppose that only one

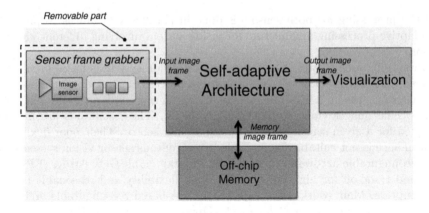

Fig. 1. Overall architecture of the vision system

sensor is active at a time while the remaining sensor are unused. The overall architecture of the vision system is presented in Fig. 1.

The system is mainly composed of a frame grabber including the sensor, the proposed self-adaptive architecture, visualization hardware and external computing resources such as off-chip memory. The Sensor frame grabber is a removable part that can be replaced when the sensor has to be changed.

3.1 Sensor Frame Grabber

Whatever is the sensor type, each Sensor frame grabber has the same hardware organization. Figure 2 depicts the sensor frame grabber organization.

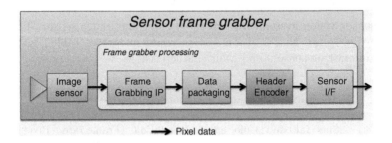

Fig. 2. Architecture of the sensor preprocessing

The image sensor is followed by the Frame Grabbing Intellectual Property (IP). This IP is specific to the image sensor and it is used to drive the latter and readout the pixel values from the imager's pixel array. Then the image frame data are gathered into packets and sent to the processing architecture by the communication core. The novelty that this work brings into the Sensor frame grabber is the Header Encoder.

This Header Encoder adds a Stream Header to the existing image stream. This Stream Header is used as an identity card of the sensor. It conveys the sensor's characteristics to the Self-adaptive architecture.

3.2 Stream Header

Packet header concept is widely used in Network-On-Chip to add a description of the packet within the packet. In real time vision application, this concept was used to add a description of the image frame into the image packets [12]. In our system, the Header Encoder is placed next to the Data packaging processing in the Sensor frame grabber. The Stream Header does not need any extra datapath. It is integrated into the image stream so that the image data and the Stream Header data use the same datapath.

The Stream Header contains information that intends to be used to adapt the image processing in the Self-adaptive architecture. The Stream Header is mainly composed of the sensor's characteristics. Its details are given in Fig. 3. In this figure, frame synchronizations are presented in blue arrow. They announces the beginning of a frame. The Stream Header in inserted between the frame synchronization and the image frame data.

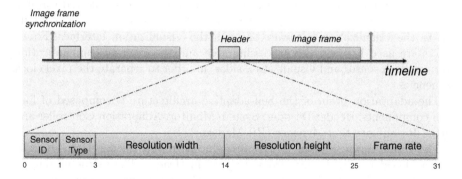

Fig. 3. Stream Header data format

The Stream Header is composed of five data. There are the sensor ID, the sensor type, the resolution separated into resolution width and resolution height and finally the frame rate. The sensor ID has two functions. In one hand, it is used to recognize the given sensor among several sensors that the vision system is ought to deal with. In other hand, the sensor ID will be used by the Self-adaptive architecture to detect when the sensor is switched. The sensor type is the main information that is used to adapt the image processing.

3.3 Self-Adaptive Architecture

Image frame data and Stream Header are then processed in the Self-adaptive architecture. Figure 4 depicts the Self-adaptive architecture and its components.

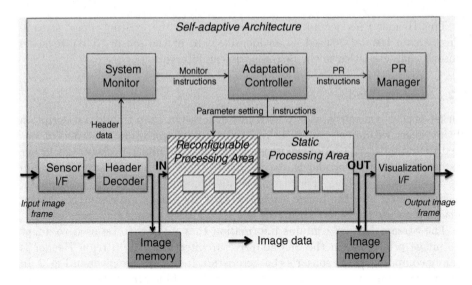

Fig. 4. Self-adaptive architecture

Input image stream coming from the Sensor frame grabber is collected through the Sensor interface (I/F), then processed in the processing area, and finally sent to the visualization hardware through the Visualization interface. Frame buffers are used between the processing area and the external communication interface both sensor and visualization sides, in order to separate the pixel clock frequencies.

The adaptation brain of the Self-adaptive architecture is composed of four main components: Header Decoder, System Monitor, Adaptation Controller and Partial Reconfiguration Manager (PR Manager).

The input image stream is first introduced in the Header Decoder. The Header Decoder extracts data from the Stream Header. The extracted data is then sent separately to the System Monitor. The output image frame from the Header Decoder does not contain the Stream Header anymore, only the image data goes through the processing area.

All the image processing are implemented in the processing area. This area is divided into two parts: Reconfigurable processing area and Static processing area. Actually, the image processing chain to process the sensor data can be divided into two parts. The first part of the processing depends on the sensor type whereas the second one is common for every sensor. Image processing chain of two different sensors differs only in the first part. Details on the considered image processing chains in this work are given in the case study in Sect. 4.

When the sensor is switched, only the first part of the image processing chain is modified. Hence, this part is placed in the Reconfigurable processing area. The adaptation of the first part consist in Dynamic and Partial Reconfiguration (DPR) of the Reconfigurable processing area. In the second part, only resolution parameters are adapted without modifying the rest of image processing chain.

3.4 Adaptation Process

The System Monitor collects Stream Header data coming from the Header Decoder. It has local registers where these data can be saved. The System Monitor first compares the previously saved sensor ID and the new sensor ID. If the two IDs are not the same, it turns a *new-sensor* flag to the active state to inform the Adaptation Controller that the sensor has been switched. Then it saves the new Stream Header data in the local registers.

The System Monitor has also an additional register to save the current sensor type of the system. This information will be used by the Adaptation Controller. Finally, the System Monitor transmits the saved data to the Adaptation Controller.

The Adaptation Controller is responsible for the adaptation of the image processing when the sensor is switched. The adaptation process follows the Finite State Machine (FSM) presented in Fig. 5.

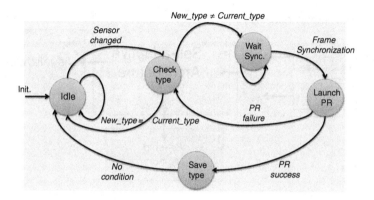

Fig. 5. Finite State Machine of the adaptation process

This FSM has five states. Initially, the system starts with the default **Idle** state. If the *new-sensor* flag is active, it means that the sensor has been switched. Consequently, the FSM goes to the **Check type** state. In this state, the current sensor's type is compared to the new sensor's type. If the new one is different than the current one, then the **Wait sync** state is reached. Otherwise, the FSM returns to **Idle** state.

In the **Wait sync** state the FSM waits for the next image frame synchronization before sending a PR request. Actually, if the Partial Reconfiguration is launched in the middle of the processing of a frame, it will corrupt the output image stream. Once an image frame synchronization has passed, the Adaptation Controller fires PR request to the PR Manager (**Launch PR**). The Adaptation Controller also gives the new sensor's type so that the PR Manager configures the right image processing chain in the Reconfigurable processing area.

When the Partial Reconfiguration has finished successfully, the System Monitor saves the new sensor's type (**Save status**) before coming back to the **Idle** state. In case of a PR failure, the FSM returns to **Check type** state in order to restart the adaptation process.

4   Experimental Prototyping

Evaluation of the proposed Self-Adaptive architecture concept has been made in a case of study with two sensors. A test bench with one color sensor and one infrared sensor has been used for experimental purpose only. Figure. 6 depicts the Self-adaptive architecture with this test bench. In this test bench, the selection of the sensor is made by a manual switch. This test bench intends to make easy the technical manipulations to switch between the two sensors, otherwise the concept of the Self-adaptive architecture remains unchanged.

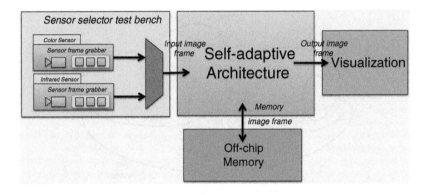

Fig. 6. Architecture of the vision system with the test bench

Image processing chain for color and infrared sensor of this case of study are given in Fig. 7. The Reconfigurable Area contains the image processing that are specific to the sensor. These processing perform image restoration.

In the case of the color sensor, these processing are White Balance, Demosaicing and finally color space transform processing to switch from RGB space to YCbCr space. In the case of the infrared sensor, the restoration processing are Non-Uniformity-Correction (NUC) and Median Filter processing.

Image quality enhancement processing are common for both color and infrared sensor so they are placed in the static area. These processing are Digital zoom and frame size readjusting, Contour Enhancing and Contrast Enhancing.

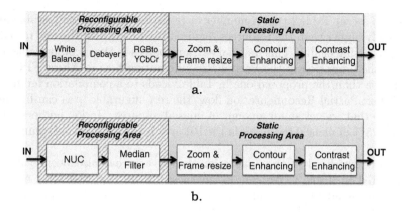

Fig. 7. Image processing chain for color(a) and infrared(b) sensor

5 Evaluation and Results

The proposed architecture was implemented in a Cyclone V Altera FPGA [13].
In FPGA implementation, area occupation of an architecture is given in terms
of logical and memory resources, which are look-up-tables(LUT), registers and
memory blocks.

This work highlights the advantage of the partial reconfiguration for resource
optimization on a recent FPGA. Instead of using two static processing areas for
the changing part of the image processing chain, resources of the same Reconfigurable processing area are temporally multiplexed for both color and infrared
sensor. The size of Reconfigurable processing area depends on the resource
requirements of the design that is deployed in the Reconfigurable processing
area. This area shall include all the resources required by anyone of the two
reconfigurable processing chains. Usually, the reconfigurable area includes little
bit more resources than the exact required amount.

Table 2 gives details about the selected reconfigurable area in the proposed
prototype. It gives available resources in the reconfigurable area. Percentage
between brackets in Table 2 represents the fraction of resources used by each
reconfigurable processing chain within the total resources available in the reconfigurable area. The selected reconfigurable area has a 16-columns width and a
51-rows height.

Table 2. Resource utilization of processing chain

	ALUT	Register	Memory (bit)
Reconfigurable area	12 240	24 480	1 044 480
Reconfigurable color processing	1 733 (14 %)	1 678 (6.9 %)	122 800 (12 %)
Reconfigurable infrared processing	1 747 (14 %)	5 055 (21 %)	98 304 (9.4 %)

We can see in Table 2 that a number of resources in the reconfigurable area is extremely higher than the resources required from anyone of the two reconfigurable processing chains. This unusual extra resources in the reconfigurable area is due to the limitation of the Partial Reconfiguration flow in Altera FPGA. A smaller area than the proposed one in Table 2 leads to a compilation failure.

In Xilinx Partial Reconfiguration flow, the reconfigurable area can be finely defined, so that we get small amount of unused resources. In [5], authors relate that only 25 % of unused resources in the PR area is enough to avoid compilation failure.

Resource overhead of the adaptation components has also been evaluated and reported in Table 3. We can see in these results that the total resource overhead of the adaptation components are quite insignificant compared to the resource utilization of the full design. This low resource overhead is justified by the modest complexity of the proposed adaptation process.

Adaptation times are given in Table 4. This table gives adaptation time in milliseconds for a system clock of 100 MHz. Most of the adaptation time is due to the partial reconfiguration of the FPGA. An average adaptation time of 75 ms has been measured both for color to infrared and infrared to color sensor switches. The Adaptation process time includes all the states of the FSM of the Adaptation Controller excluding the partial reconfiguration time.

Partial bitstreams of the two Reconfigurable processing areas has not the same size. The color partial bitstream is about 5.8 MB whereas the infrared one is about 5.7 MB. As the two partial bitstreams has not the same size, the reconfiguration time differs lightly between the two adaptations.

Table 3. Resource overhead of adaptation components

	ALUT	Register	Memory (bit)
Full design	24 947	31 945	1 726 056
Header encoder	43	5	0
Header decoder	12	24	0
System monitor	8	28	0
Adaptation controller	31	22	0
PR manager	22	15	0
Total adaptation components	**116 (0,5 %)**	**94 (0,3 %)**	**0**

Table 4. Adaptation times

	Time (ms)
Header decoding	0.00003
Adaptation process	0.00015
PR color-to-infrared	75.02
PR infrared-to-color	75.10

A partial reconfiguration time of 75 ms is to long compared to usual times in Xilinx FPGA-based designs. This time represents about two image frames time in a 25 fps system. This unusual long time is due to the limitation of the partial reconfiguration in recent Altera FPGA. Actually, Altera FPGAs of the series V offers only a column-wise partial reconfiguration like the first Virtex-II Xilinx FPGAs [14]. However, two frames time is acceptable for sensor switch in non-constrained applications. Future works will focus on implementation of this architecture in a recent Xilinx FPGA enabling partial reconfiguration such as series 7 FPGAs, to get better reconfiguration performance.

6   Conclusion

In this paper, we have presented a self-adaptive architecture for multi-sensor vision system. The novelty of this work is that it proposes a new adaptive processing architecture which can deal with multiple and heterogeneous sensors. This architecture can dynamically adapt itself as a consequence of on-the-fly sensor switch. This self-adaptive architecture is based on a stream header concept and an adaptation controller to make the system aware of the sensor switch and hence to adapt the processing architecture.

This work offers a performance evaluation of the proposed concept in an FPGA implementation. The architecture was implemented in a Cyclone V Altera FPGA. Processing architecture is dynamically adapted by Dynamic and Partial Reconfiguration feature of FPGA. Rather than a performance comparison, the aim of this work is to give the proof of concept above all.

An average adaptation time of 75 ms has been measured. This time is mostly representative of the partial reconfiguration time. Because of the limitation of the Altera's newly released partial reconfiguration technology, reconfiguration times are high. Nevertheless, this time remains admissible for non-constrained applications. Resource overhead due to self-adaptation is almost insignificant compared to the resources utilization of the full design. The concept of this architecture is a promising start for further work on self-adaptive vision systems.

This work will be extended for multi-sensor vision systems with multiple and parallel streams, such as color-infrared image fusion systems. Future works will focus on improvement of the monitoring solution to enable a full self-awareness and environment-awareness to the vision system.

References

1. Nakamura, J.: Image Sensors and Signal Processing for Digital Still Cameras, pp. 143–178. CRC Press Taylor & Francis Group, Boca Raton (2005). Chapter 5, CMOS Image Sensors
2. Platzner, M., Teich, J., Wehn, N.: Dynamically Reconfigurable Systems, Architectures, Design Methods and Applications, pp. 375–415. Springer, The Netherlands (2010). Chapter 18
3. Birem, M., Berry, F.: DreamCam :a modular FPGA-based smart camera architecture. J. Syst. Archit. **60**, 519–527 (2014)

4. Raikovich, T., Fehér, B.: Application of partial reconfiguration of FPGAs in image processing Conference on Ph.D. Research in Microelectronics and Electronics, (2010)
5. Khalifat, J., Arslan, T.: A novel dynamic partial reconfiguration design for automatic white balance. In: NASA/ESA Conference on Adaptive Hardware and Systems (AHS), Leicester, July 2014
6. van der Horst, J., van Leeuwen, R., Broers, H., Kleihorst, R., Jonker, P.: A real-time stereo SmartCam, using FPGA, SIMD and VLIW. In: Proceedings of 2nd Workshop on Applications of Computer Vision, pp. 1–8 (2006)
7. Muscoloni, A., Mattoccia, S.: Real-time tracking with an embedded 3D camera with FPGA processing. In: International Conference on 3D Imaging (IC3D), Liège, December 2014
8. Texas Instruments. DaVinci Video Processors and Digital Media System-On-Chip. tms320dm6446 Datasheet, 30 September 2010
9. ON Semiconductor, formerly Aptina Imaging. 1/3-Inch CMOS Digital Image Sensor. AR0130 Datasheet, 11/2014
10. E2V. 1.3 Mpixels B&W and Color CMOS image Sensor. E2V Datasheet, 10/2011
11. ULIS. Pico640 Gen2. Ulis Datasheet, 01/2015
12. Ngan, N., Dokladalova, E., Akil, M.: Dynamically adaptable NoC router architecture for multiple pixel streams applications. In: IEEE International Symposium on Circuits and Systems (ISCAS'12), Seoul, May 2012
13. Altera: Cyclone V Device Overview. CV-51001, 2015–06-12
14. Bhandari, S., Subbaraman, S., Pujari, S., Cancare, F., Bruschi, F., Santambrogio, M.D., Grassi, P.R.: High speed dynamic partial reconguration for real time multimedia signal processing. In: 15th Euromicro Conference on Digital System Design, Izmir (2012)

Exceptional Configurations of Quantum Walks with Grover's Coin

Nikolajs Nahimovs and Alexander Rivosh[(⊠)]

Faculty of Computing, University of Latvia, Raina Bulv. 19, Riga 1586, Latvia
{nikolajs.nahimovs,alexander.rivosh}@lu.lv

Abstract. We study search by quantum walk on a two-dimensional grid using the algorithm of Ambainis, Kempe and Rivosh [AKR05]. We show what the most natural coin transformation — Grover's diffusion transformation — has a wide class of exceptional configurations of marked locations, for which the probability of finding any of the marked locations does not grow over time. This extends the class of known exceptional configurations; until now the only known such configuration was the "diagonal construction" by [AR08].

1 Introduction

Quantum walks are the quantum counterparts of classical random walks [Por13]. They have been useful for designing quantum algorithms for a variety of problems [CC+03, AKR05, MSS05, BS06, Amb07]. In many of those applications, quantum walks are used as a tool for search.

To solve a search problem using quantum walks, we introduce the notion of marked locations. Marked locations correspond to elements of the search space that we want to find. We then perform a quantum walk on the search space with one transition rule at the unmarked locations, and another transition rule at the marked locations. If this process is set up properly, it leads to a quantum state in which marked locations have higher probability than the unmarked ones. This state can then be measured, finding a marked location with a sufficiently high probability. This method of search using quantum walks was first introduced in [SKW03] and has been used many times since then.

We study search by quantum walk on a finite two-dimensional grid using the algorithm of Ambainis, Kempe and Rivosh (AKR). The original [AKR05] paper proves that after $O(\sqrt{N \log N})$ steps, a quantum walk with one or two marked locations reaches a state that is significantly different from the initial state. Szegedy [Sze04] has generalized this to an arbitrary number of marked locations. Thus, quantum walks can detect the presence of an arbitrary number of marked locations. [AKR05] also shows that for one or two marked locations, the probability of finding a marked location after $O(\sqrt{N \log N})$ steps is $O(1/\log N)$. Thus, for one or two marked locations, the AKR algorithm can also find a

N. Nahimovs is supported by EU FP7 project QALGO, A. Rivosh is supported by ERC project MQC.

J. Kofroň and T. Vojnar (Eds.): MEMICS 2015, LNCS 9548, pp. 79–92, 2016.
DOI: 10.1007/978-3-319-29817-7_8

marked location. For a larger number of marked locations, this is not always the case. Ambainis and Rivosh [AR08] have found an exceptional configuration of marked locations for which AKR algorithm fails to find any of marked locations.

A step of the AKR algorithm consists of two transformations: the coin-flip transformation, which acts on internal state of the walker and rearranges the amplitudes of going to adjacent locations, and the shift transformation, which moves the walker between the adjacent locations. The original AKR algorithm uses D – Grover's diffusion transformation – as the coin transformation for the unmarked locations and $-I$ as the coin transformation for the marked locations[1]. Another natural choice for the coin transformation is D for the unmarked locations and $-D$ for the marked locations.

Nahimovs and Rivosh [NR15] has analysed the dependence of the running time of the AKR algorithm on the number and placement of marked locations and showed that the algorithm is inefficient for grouped marked locations (multiple marked locations placed near-by). They showed that for a $k \times k$ group of marked locations, the AKR algorithm needs the same number of steps and has the same probability to find a marked location as for $4(k-1)$ marked locations placed as the perimeter of the group (with all internal locations being unmarked). The reason for the inefficiency is the coin transformation used by the original AKR algorithm. The original coin transformation does not rearrange direction amplitudes within a marked location. As a result, marked locations inside the group have almost no effect on the number of steps and the probability to find a marked location of the algorithm.

We try to solve the above problem by replacing the original coin transformation by one which rearranges amplitudes within a marked location. We use the most natural choice of such coin transformation — Grover's diffusion transformation. We show what while the modified algorithm works well for some of the "problematic" configurations, it has a wide class of exceptional configurations of marked locations, for which the probability to find any of marked locations does not grow over time. Namely, we prove that any block of marked locations of size $2m \times l$ or $m \times 2l$, that is the block with one of its sides consisting of even number of marked locations, is the exceptional configuration. This extends the class of known exceptional configurations; until now the only known such configuration was the "diagonal construction" by [AR08].

The AKR algorithm is very generic and can be adapted to other types of graphs. In the appendix we describe the AKR algorithm for general graphs and generalize the exceptional configurations that have been found.

2   Quantum Walks in Two Dimensions

Suppose we have N items arranged on a two dimensional grid of size $\sqrt{N} \times \sqrt{N}$. We denote $n = \sqrt{N}$. The locations on the grid are labelled by their x and y coordinates as (x, y) for $x, y \in \{0, \ldots, n-1\}$. We assume that the grid has

---

[1] According to authors of [AKR05], this coin transformation was chosen because it leads to a simpler proof.

periodic boundary conditions. For example, going right from a location $(n-1, y)$ on the right edge of the grid leads to the location $(0, y)$ on the left edge of the grid.

To introduce a quantum version of a random walk, we define a location register with basis states $|i, j\rangle$ for $i, j \in \{0, \ldots, n-1\}$. Additionally, to allow non-trivial walks, we define a direction or coin register with four basis states, one for each direction: $|\Uparrow\rangle$, $|\Downarrow\rangle$, $|\Leftarrow\rangle$ and $|\Rightarrow\rangle$. Thus, the basis states of the quantum walk are $|i, j, d\rangle$ for $i, j \in \{0, \ldots, n-1\}$ and $d \in \{\Uparrow, \Downarrow, \Leftarrow, \Rightarrow\}$. The state of the quantum walk is given by:

$$|\psi(t)\rangle = \sum_{i,j} (\alpha_{i,j,\Uparrow}|i, j, \Uparrow\rangle + \alpha_{i,j,\Downarrow}|i, j, \Downarrow\rangle + \alpha_{i,j,\Leftarrow}|i, j, \Leftarrow\rangle + \alpha_{i,j,\Rightarrow}|i, j, \Rightarrow\rangle).$$

A step of the quantum walk is performed by first applying $I \otimes C$, where C is unitary transform on the coin register. The most often used transformation on the coin register is the Grover's diffusion transformation D:

$$D = \frac{1}{2} \begin{pmatrix} -1 & 1 & 1 & 1 \\ 1 & -1 & 1 & 1 \\ 1 & 1 & -1 & 1 \\ 1 & 1 & 1 & -1 \end{pmatrix}.$$

Then, we apply the shift transformation S:

$$\begin{aligned} |i, j, \Uparrow\rangle &\rightarrow |i, j-1, \Downarrow\rangle \\ |i, j, \Downarrow\rangle &\rightarrow |i, j+1, \Uparrow\rangle \\ |i, j, \Leftarrow\rangle &\rightarrow |i-1, j, \Rightarrow\rangle \\ |i, j, \Rightarrow\rangle &\rightarrow |i+1, j, \Leftarrow\rangle \end{aligned}$$

Notice that after moving to an adjacent location we change the value of the direction register to the opposite. This is necessary for the quantum walk algorithm of [AKR05] to work.

We start the quantum walk in the state

$$|\psi_0\rangle = \frac{1}{\sqrt{4N}} \sum_{i,j} \left(|i, j, \Uparrow\rangle + |i, j, \Downarrow\rangle + |i, j, \Leftarrow\rangle + |i, j, \Rightarrow\rangle \right).$$

It can be easily verified that the state of the walk stays unchanged, regardless of the number of steps.

To use the quantum walk as a tool for search, we mark some locations. For the unmarked locations, we apply the same transformations as above. For the marked locations, we apply $-I$ instead of D as the coin flip transformation. The shift transformation remains the same for both the marked and the unmarked locations.

Another way to look at the step of the algorithm is that we first perform a query Q transformation, which flips signs of amplitudes of marked locations, then conditionally perform the coin transformation (I or D depending on whether the location is marked or not) and then perform the shift transformation S. In the case of the modified coin (D for unmarked locations and $-D$ for marked locations), the step of the algorithm consists of the query Q followed by D followed by S.

If there are marked locations, the state of the algorithm starts to deviate from $|\psi(0)\rangle$. It has been shown [AKR05] that after $O(\sqrt{N \log N})$ steps, the inner product $\langle\psi(t)|\psi(0)\rangle$ becomes close to 0.

In the case of one or two marked locations, the AKR algorithm finds a marked location with $O(1/\log N)$ probability. The probability is small, thus, the algorithm uses amplitude amplification to get $\Theta(1)$ probability. The amplitude amplification adds an additional $O(\sqrt{\log N})$ factor to the number of steps. Thus, the total running time of the algorithm is $O(\sqrt{N} \log N)$.

3   Quantum Walks with Grover's Coin

The coin transformation used by the AKR algorithm does not rearrange amplitudes within a marked location. As it was shown in [NR15], this results in the AKR algorithm being inefficient for grouped marked locations.

In this section we consider an alternative coin transformation which rearranges amplitudes at both the marked and unmarked locations. As the most natural choice of such transformation we use D and $-D$ as coin for the unmarked and marked locations, respectively. We refer this coin transformation as Grover's coin and the original coin transformation of the AKR algorithm as the AKR coin.

First, we compare the Grover and AKR coins for a $\sqrt{k} \times \sqrt{k}$ group of marked locations ("inefficient" configuration of [NR15]). We run a series of numerical experiments and demonstrate that in some cases, Grover's coin works better than AKR coin.

Next, we show a wide class of exceptional configurations of marked locations, for which the probability to find any of marked locations does not grow over time. We explain exceptional configurations based on stationary states of a step of the algorithm with Grover's coin.

3.1   AKR Vs Grover's Coin: Numerical Experiment Results

In this subsection, we compare the AKR algorithm with the Grover and AKR coins. We consider k marked locations placed as a $\sqrt{k} \times \sqrt{k}$ square and compare the number of steps and the probability to find a marked location for $\sqrt{k} \in [2, \ldots, 10]$ and grid sizes from 100×100 to 1000×1000 with step 100.

Table 1 shows the results of numerical simulations for $k = 9$ (3×3 group of marked locations). As one can see, the algorithm with Grover coin needs more steps, however, it has much higher probability of finding a marked location and, thus, has smaller total running time (number of steps of the single run of the algorithm divided by square root of the probability).

Table 2 shows the ratio between running times of the algorithm with the AKR and Grover coins for $k = 9$. Table 3 shows the ratio for different k and N. As one

Table 1. Number of steps, probability and running time for the algorithm with the AKR and Grover coins for $k = 9$ and different N.

Grid size	AKR coin			Grover's coin		
	Steps	Probability	Runtime	Steps	Probability	Runtime
100	156	0.086454	531	318	0.556187	427
200	345	0.066591	1337	653	0.527665	899
300	544	0.063212	2164	993	0.510679	1390
400	749	0.058022	3110	1337	0.499213	1893
500	959	0.055813	4060	1685	0.49053	2406
600	1172	0.055086	4994	2034	0.483683	2925
700	1389	0.052851	6042	2386	0.478038	3451
800	1608	0.051962	7055	2739	0.473336	3982
900	1829	0.049888	8189	3093	0.469256	4516
1000	2052	0.049255	9246	3449	0.465662	5055

can see, the ratio between the running times decreases with k and increases with N. The below results are obtained by running a series of numerical simulations. Thus, the interesting and important open question here is to find analytical formula giving the running time of the algorithm with AKR and Grover's coins for a group of marked locations.

Table 2. Ratio between running times for the AKR and Grover coins for $k = 9$ and different N.

Grid size	AKR coin	Grover's coin	Ratio
100	531	427	1.243559719
200	1337	899	1.487208009
300	2164	1390	1.556834532
400	3110	1893	1.642894876
500	4060	2406	1.687448047
600	4994	2925	1.707350427
700	6042	3451	1.75079687
800	7055	3982	1.771722752
900	8189	4516	1.813330381
1000	9246	5055	1.829080119

For $k = 4$, the quantum walk with Grover's coin does not find any of the marked locations. More precisely, the overlap between the current and initial state of the algorithm never reaches 0, but stays close to 1. Thus, the probability

Table 3. The ratio between the running times for the AKR and Grover coins for different k and N.

Grid size	k = 9	k = 25	k = 49	k = 81
100	1.243559719	1.016453382	0.771014493	0.624553039
200	1.487208009	1.286351472	1.059413028	0.829787234
300	1.556834532	1.420867526	1.205965909	1.02627451
400	1.642894876	1.480191554	1.268619838	1.123094959
500	1.687448047	1.552176918	1.345050619	1.196541248
600	1.707350427	1.631473534	1.406490777	1.224640497
700	1.75079687	1.655281776	1.458191978	1.275856335
800	1.771722752	1.695495113	1.500870777	1.344321812
900	1.813330381	1.730009407	1.56015444	1.356277391
1000	1.829080119	1.775771891	1.591205438	1.411492122

to find a marked location does not grow with the number of steps. The same holds for $k = 16$, $k = 36$, $k = 64$, etc., that is, for any k having even $\sqrt{k}$. Moreover, the same effect holds for any block of marked locations of size $2m \times l$ and $m \times 2l$, that is, the block with one of it sides consisting of an even number of marked locations.

Therefore, while the algorithm with Grover's coin has a smaller running time, compared to the algorithm with the AKR coin, for some configurations, it completely fails for other configurations.

3.2 Exceptional Configurations of a Quantum Walk with Grover's Coin

As it was mentioned in the previous subsection, the AKR algorithm using Grover's coin fails to find any group of marked locations of size $2m \times l$ or $m \times 2l$. In this subsection, we explain this phenomenon. First, we prove that a group of marked locations of size 1×2 (and by symmetry 2×1) is an exceptional configuration. Next, we show how one can extend the argument to any group of size $2m \times l$ or $m \times 2l$.

Consider a grid of size $\sqrt{N} \times \sqrt{N}$ with two marked locations (i, j) and $(i, j + 1)$. Let $|\phi_{stat}^a\rangle$ be a state having amplitudes of all basis states except $|i, j, \Rightarrow\rangle$ and $|i, j+1, \Leftarrow\rangle$ equal to a and amplitudes of basis states $|i, j, \Rightarrow\rangle$ and $|i, j+1, \Leftarrow\rangle$ equal to $-3a$ (see Fig. 1). Then this state is not changed by a step of the algorithm.

Theorem 1. Let locations (i, j) and $(i, j + 1)$ be marked and let

$$|\phi_{stat}^a\rangle = \sum_{i,j,d} a|i, j, d\rangle - 4a|i, j, \Rightarrow\rangle - 4a|i, j + 1, \Leftarrow\rangle.$$

Then, $|\phi_{stat}^a\rangle$ is not changed by a step of the algorithm with Grover's coin.

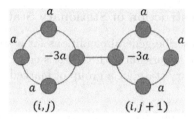

Fig. 1. Stationary state for 1×2 block of marked locations.

Proof. Consider the effect of a step of the algorithm on $|\phi^a_{stat}\rangle$. The query transformation changes the signs of all the amplitudes of the marked locations. The coin transformation perform an inversion above the average: for non-marked locations, it does nothing as all amplitudes are equal to a; for marked locations, the average is 0, so the inversion results in sign flip. Thus, CQ does nothing for amplitudes of non-marked locations and twice flips the sign of amplitudes of marked locations. Therefore, we have

$$CQ|\phi^a_{stat}\rangle = |\phi^a_{stat}\rangle.$$

The shift transformation swaps the amplitudes of near-by locations. For $|\phi^a_{stat}\rangle$, it swaps a with a and $-3a$ with $-3a$. Thus, we have

$$SCQ|\phi^a_{stat}\rangle = |\phi^a_{stat}\rangle.$$

$\square$

Consider the initial state of the algorithm

$$|\psi_0\rangle = \frac{1}{\sqrt{4N}} \sum_{i,j} (|i,j,\Uparrow\rangle + |i,j,\Downarrow\rangle + |i,j,\Leftarrow\rangle + |i,j,\Rightarrow\rangle).$$

It can be written as

$$|\psi_0\rangle = |\phi^a_{stat}\rangle + 4a(|i,j,\Rightarrow\rangle + |i,j+1,\Leftarrow\rangle),$$

for $a = 1/\sqrt{4N}$. Therefore, the only part of the initial state which is changed by the step of the algorithm is

$$\sqrt{\frac{4}{N}}(|i,j,\Rightarrow\rangle + |i,j+1,\Leftarrow\rangle).$$

Now, consider a group of marked locations of size $m \times 2l$. It is equivalent to $m \times l$ groups of marked locations of size 1×2. Thus, by repeating the above construction $m \times l$ times, one can build the stationary state for the group. Moreover, if $m > 1$, then the group of size $2m \times l$ has multiple tilings by groups of size 2×1 and 1×2, where each tiling corresponds to a stationary state of the step of the algorithm.

3.3 Alternative Construction of Stationary States

In this subsection we describe general conditions for a state to be a stationary state of the step of ARK algorithm with Grover's coin. and give an alternative construction of a stationary state for a group of marked locations.

General Conditions. A stationary state from the previous section has three properties:

1. All directional amplitudes of unmarked locations are equal. This is necessary for the coin transformation to have no effect on the unmarked locations.
2. The sum of the directional amplitudes of any marked location is equal to 0. This is necessary for the coin transformation to have no effect on marked locations.
3. Direction amplitudes of two adjacent locations pointing to each other are equal. This is necessary for the shift transformation to have no effect on the state.

It is easy to see that any state having these three properties is not changed by the step of AKR algorithm with Grover's coin and, thus, is a stationary state.

Alternative Construction of a Stationary State. Consider a group of marked locations of size $m \times l$. Without the loss of generality, let $m \leq l$. We build the stationary state iteratively. First, we set all directional amplitudes of the unmarked locations to a. Next, we iteratively set amplitudes of the marked locations. On each iteration we set the amplitudes of one rectangular layer of the marked locations, starting from the outer layer (the perimeter of the group). The iteration is as follows:

1. Set two directional amplitudes of a location pointing to its perimeter-wise neighbours to $-a$.
2. Set two other directional amplitudes of the location (pointing to the inner and the outer layers) to a.

Figure 2 illustrates the first iteration of the construction for the group of marked locations of size 4×5. Amplitudes set on step 1 are colored in blue. Amplitudes set on step 2 are colored in red. Figure 3 shows the resulting stationary state after all amplitudes are set.

The iteration reduces the size of the unprocessed group of marked locations from $m \times l$ to $m' \times l'$, where $m' = m - 2$ and $l' = l - 2$. We repeat the iteration while $m' \geq 2$. If $m' = 0$, we have assigned values to all direction amplitudes and, thus, have built a stationary state. If $m' = 1$, there are three possibilities:

- $m' = l' = 1$. The construction is not possible. The initial group of marked locations was of odd-times-odd size.
- $m' = 1$, $l' > 1$, l is odd. The construction is not possible. The initial group of marked locations was of odd-times-odd size.

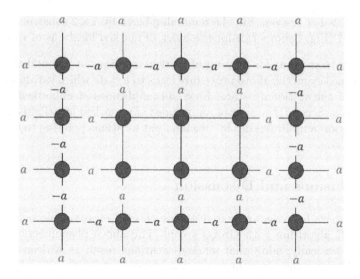

Fig. 2. The first iteration for a group of marked locations of size 4×5.

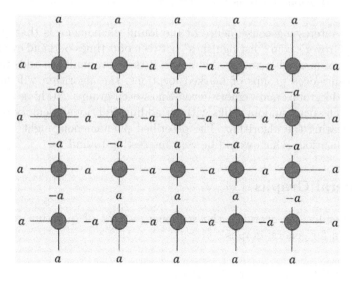

Fig. 3. The stationary state for a group of marked locations of size 4×5.

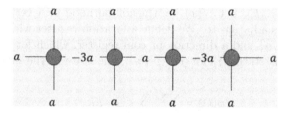

Fig. 4. The stationary state for a group of marked locations of size 4×5.

– $m' = 1$, $l' > 1$, l' is even. Fill the remaining block by 1×2 constructions from Theorem 1 (Fig. 4 shows this for the block of marked locations of size 1×4.).

It is easy to see that for a group of marked locations of size odd-times-even and even-times-even, the above procedure leads to a state which satisfies all three properties of the stationary state. First, all amplitudes of unmarked locations are equal to a. Second, the sum of amplitudes of a marked location is always 0. Third, direction amplitudes of any two adjacent locations pointing to each other are equal.

4   Conclusions and Discussion

In this paper we have demonstrated a wide class of exceptional configurations for the AKR algorithm with Grover's coin. The above phenomenon is purely quantum. Classically, additional marked locations result in a decrease of the number of steps of the algorithm and an increase of the probability of finding a marked location. Quantumly, as we have demonstrated in the paper, the addition of a marked location can drastically drop the probability of finding a marked location.

Another interesting consequence of the found phenomena is that the algorithm with Grover's coin "distinguishes" between odd-times-odd and even-times-even groups of marked locations. Moreover, if there are multiple odd-times-odd and even-times-even groups of marked locations, the algorithm will find only odd-times-odd groups and "ignore" even-times-even groups. Nothing like this is possible for classical random walks without adding additional memory resources and complicating the algorithm. The described phenomenon might have algorithmic applications which would be very interesting to find.

A   General Graphs

In this appendix, we overview the stationary states of quantum walks with Grover's coin for general graphs.

Quantum Walks on a General Graph

Consider a graph $G = (V, E)$ with a set of vertices V and a set of edges E. Let $n = |V|$ and $m = |E|$. Let $N(x)$ be a neighbourhood of a vertex x, that is a set of vertices x is adjacent to. We define a location register with n basis states $|i\rangle$ for $i \in \{1, \ldots, n\}$ and a direction or coin register, which for a vertex v_i has $d_i = \deg(v_i)$ basis states $|j\rangle$ for $j \in N(v_i)$. The state of the quantum walk is given by:

$$|\psi(t)\rangle = \sum_{i=1}^{n} \sum_{j \in N(v_i)} \alpha_{i,j} |i, j\rangle.$$

A step of the quantum walk is performed by first applying $I \otimes C$, where C is a unitary transformation on the coin register. The usual choice of transformation on the coin register is Grover's diffusion transformation D. Then, we apply the shift transformation S:

$$S = \sum_{i=1}^{n} \sum_{j \in N(v_i)} |j, i\rangle\langle i, j|,$$

which for each pair of connected vertices i, j swaps an amplitude of vertex i pointing to j with an amplitude of vertex j pointing to i.

We start the quantum walk in the equal superposition over all pairs vertex-direction:

$$|\psi_0\rangle = \frac{1}{\sqrt{n \cdot \deg(G)}} \sum_{i=1}^{n} \sum_{j \in N(v_i)} |i, j\rangle,$$

where $\deg(G) = \sum_i \deg(v_i)$. It can be easily verified that the state of the walk stays unchanged, regardless of the number of steps.

To use the quantum walk as a tool for search, we mark some vertices. For the unmarked vertices, we apply the same transformations as above. For the marked vertices, we apply $-I$ instead of D as the coin flip transformation. The shift transformation remains the same for both the marked and unmarked vertices.

Another way to look at a step of the algorithm is that we first perform a query Q transformation, which flips signs of amplitudes of marked vertices, then conditionally perform the coin transformation (I or D depending on whether a vertex is marked or not) and then perform the shift transformation S. In case of the Grover's coin the step of the algorithm is the query Q followed by D followed by S.

Stationary States of the Quantum Walk with Grover's Coin for General Graphs

Consider a graph $G = (V, E)$ with two marked vertices v_i and v_j. Let vertices be connected and let each of them be connected to some other k vertices. Let $|\phi_{stat}^a\rangle$ be a state having amplitudes of all basis states except $|i, j\rangle$ and $|j, i\rangle$ equal to a and amplitudes of basis states $|i, j\rangle$ and $|j, i\rangle$ equal to $-ka$ (see Fig. 5). Then this state is not changed by a step of the algorithm with Grover's coin.

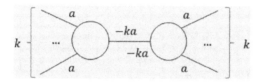

Fig. 5. Symmetric stationary state for 2 marked vertices.

Theorem 2. *Let $G = (V, E)$ be a graph with two marked vertices i and j; let $(v_i, v_j) \in E$ and $N(v_i) = N(v_j) = k + 1$; and let*

$$|\phi^a_{stat}\rangle = \sum_{i=1}^{n} \sum_{j \in N(v_i)} |i, j\rangle - (k+1)a(|i, j\rangle - |j, i\rangle).$$

Then, $|\phi^a_{stat}\rangle$ is an eigenstate of a step of the quantum walk on G with Grover's coin.

Proof. Consider the effect of a step of the algorithm on $|\phi^a_{stat}\rangle$. The query transformation changes the signs of all amplitudes of the marked vertices. The coin flip performs an inversion above the average: for unmarked vertices it does nothing as all amplitudes are equal to a; for marked vertices the average is 0, so the inversion results in sign flip. Thus, CQ does nothing for amplitudes of the unmarked vertices and twice flips the sign of amplitudes of the marked vertices. Therefore, we have

$$CQ|\phi^a_{stat}\rangle = |\phi^a_{stat}\rangle.$$

The shift transformation swaps amplitudes of adjacent vertices. For $|\phi^a_{stat}\rangle$, it swaps a with a and $-ka$ with $-ka$. Thus, we have

$$SCQ|\phi^a_{stat}\rangle = |\phi^a_{stat}\rangle.$$

$\square$

The initial state of the algorithm $|\psi_0\rangle$ can be written as

$$|\psi_0\rangle = \phi^a_{stat} + (k+1)a(|i, j\rangle + |j, i\rangle),$$

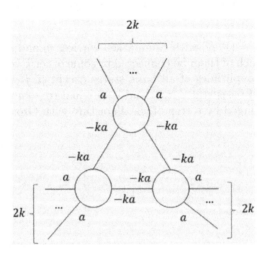

Fig. 6. Symmetric stationary state for 3 marked vertices.

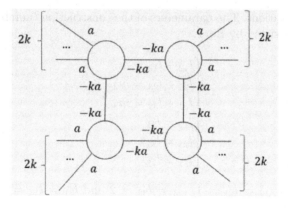

Fig. 7. Symmetric stationary state for 4 marked vertices.

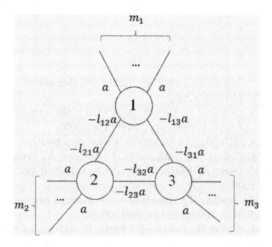

Fig. 8. Generic stationary state for 3 marked vertices.

for $a = 1/\sqrt{n \cdot \deg(G)}$. Therefore, the only part of the initial state, which is changed by a step of the algorithm, is

$$\frac{k+1}{\sqrt{n \cdot \deg(G)}}(|i,j\rangle + |j,i\rangle).$$

Next figures show similar constructions for three (Fig. 6) and four (Fig. 7) marked vertices. We give them without a proof (which is similar to the two marked vertex case). It is easy to see how one can extend the construction to any number of marked vertices.

The above constructions are symmetric in the sense that each of the marked vertices has the same number of neighbours. One can also construct a stationary state without this restriction. The Fig. 8 shows the general stationary state of three marked locations. The parameters of the construction (number of adjacent vertices) are restricted by Eq. 1.

$$
\begin{cases}
l_{12} + l_{12} = m_1 \\
l_{21} + l_{23} = m_2 \\
l_{31} + l_{32} = m_3 \\
l_{12} = l_{21} \\
l_{23} = l_{32} \\
l_{31} = l_{13}
\end{cases}
\tag{1}
$$

For example, for $l_{12} = l_{21} = 1$, $l_{23} = l_{32} = 2$ and $l_{31} = l_{13} = 3$ we will have $m_1 = 4$, $m_2 = 3$ and $m_3 = 5$.

Again, it is easy to see how one can extend the construction to any number of marked vertices.

References

[Amb07] Ambainis, A.: Quantum walk algorithm for element distinctness. SIAM J. Comput. **37**, 210–239 (2007)

[AKR05] Ambainis, A., Kempe, J., Rivosh, A.: Coins make quantum walks faster. In: Proceedings of SODA 2005, pp. 1099–1108 (2005)

[AR08] Ambainis, A., Rivosh, A.: Quantum walks with multiple or moving marked locations. In: Geffert, V., Karhumäki, J., Bertoni, A., Preneel, B., Návrat, P., Bieliková, M. (eds.) SOFSEM 2008. LNCS, vol. 4910, pp. 485–496. Springer, Heidelberg (2008)

[BS06] Buhrman, H., Spalek, R.: Quantum verification of matrix products. In: Proceedings of SODA 2006, pp. 880–889 (2006)

[CC+03] Childs, A.M., Cleve, R., Deotto, E., Farhi, E., Gutmann, S., Spielman, D.A.: Exponential algorithmic speedup by a quantum walk. In: Proceedings of 35th ACM STOC, pp. 59–68 (2003)

[MSS05] Magniez, F., Santha, M., Szegedy, M.: An $O(n^{1.3})$ quantum algorithm for the triangle problem. In: Proceedings of SODA 2005, pp. 413–424 (2005)

[NR15] Nahimovs, N., Rivosh, A.: Quantum walks on two-dimensional grids with multiple marked locations. arXiv:1507.03788 (2015)

[Por13] Portugal, R.: Quantum Walks and Search Algorithms. Springer, Heidelberg (2013)

[SKW03] Shenvi, N., Kempe, J., Whaley, K.B.: A quantum random walk search algorithm. Phys. Rev. A **67**(5), 052307 (2003)

[Sze04] Szegedy, M.: Quantum speed-up of Markov Chain based algorithms. In: Proceedings of FOCS 2004, pp. 32–41 (2004)

Performance Analysis of Distributed Stream Processing Applications Through Colored Petri Nets

Filip Nalepa[✉], Michal Batko, and Pavel Zezula

Faculty of Informatics, Masaryk University, Brno, Czech Republic
f.nalepa@gmail.com

Abstract. Nowadays, a lot of data are produced every second and they need to be processed immediately. Processing such unbounded streams of data is often run in a distributed environment in order to achieve high throughput. The challenge is the ability to predict the performance-related characteristics of such applications. Knowledge of these properties is essential for decisions about the amount of needed computational resources, how the computations should be spread in the distributed environment, etc.

In this paper, we present performance analysis of distributed stream processing applications using Colored Petri Nets (CPNs). We extend our previously proposed model with processing strategies which are used to specify performance effects when multiple tasks are placed on the same resource. We also show a detailed conversion of the whole proposed model to the CPNs. The conversion is validated through simulations of the CPNs which are compared to real streaming applications.

Keywords: Stream processing · Performance analysis · Data stream model · Colored Petri Nets

1 Introduction

1.1 Motivation

Nowadays, a lot of data are produced every second. Also such a huge amount of data often need to be reprocessed later on. There are two basic ways of doing so. The data can be processed in batches, i.e., the data are stored at first, and then the whole dataset is processed. However, there are scenarios when the data need to be processed immediately as soon as they are acquired, e.g., analyzing surveillance video footage. For such types of applications, the so called stream processing is appropriate.

The base of a stream processing application is a set of tasks (atomic computation units) that are linked by precedence constraints. This is often called workflow [2]. The tasks are used to process data streams (potentially infinite sequences of data items) which enter the application. In order to achieve high

© Springer International Publishing Switzerland 2016
J. Kofroň and T. Vojnar (Eds.): MEMICS 2015, LNCS 9548, pp. 93–106, 2016.
DOI: 10.1007/978-3-319-29817-7_9

throughput, the applications can be deployed in a distributed environment. In such cases, the tasks are placed on individual computational resources, and the data streams are sent between the resources so that the required operations can be evaluated.

For example, let there be a stream of images uploaded to a social network. Suppose we need to extract visual features of the images, classify the images, annotate them, or detect faces in them. Some of these tasks can be performed independently of each other, but for some of them, precedence constraints have to be employed. For instance, features of an image must be extracted before the image is annotated.

When building streaming applications, performance metrics (such as delay or throughput) of the final system are often big concern. For instance, one may require that each data item can be fully processed in five seconds, i.e., the maximum acceptable delay of the whole process is five seconds. Whether or not the system is able to meet such criteria is not always obvious in complex applications. The performance metrics heavily depend on the number of allocated resources and on the way how the tasks are spread throughout the network. In case of multiple tasks at a single resource, the task prioritization policy has a considerable impact on the performance too. For instance, the delays may be improved by prioritizing costly tasks so that they can keep up with the incoming data items.

It may be too late to start measuring the performance characteristics once the application has been deployed in a distributed environment since it can be difficult or expensive to deal with load balancing then. A more appropriate time to be interested in the performance is before the deployment when decisions about the required resources and tasks assignment to the resources are made.

Therefore, it is important to be able to carry out performance analysis of streaming applications without the need to deploy them. The performance analysis can serve to derive characterictics of different settings of the system, e.g., different numbers of used resources, different task placements, or different arrival rates of data items in the streams. Considering all these aspects is a key to successful planning of distributed stream processing applications.

1.2   Objectives

In this paper, we present performance analysis of distributed stream processing applications using Colored Petri Nets (CPNs) [7]. We focus mainly on multimedia data streams and on detecting various events in them. We extend our previously proposed model with processing strategy which is especially useful when there are multiple tasks placed on the same resource. We also show a detailed conversion of the whole proposed model to the CPNs. The conversion is validated through simulations of the CPNs which are compared to real streaming applications.

1.3 Related Work

A lot of work has been devoted to models and performance analysis of distributed computing systems.

Targeted at streaming applications in embedded systems, several formalisms have been proposed ([3–5]). They all work with irregular arrival patterns of data streams which cannot be described using standard periodic or sporadic event models. They make use of the arrival function for a description of a variable number of data items that can arrive to a particular component of the system each time unit. This approach allows to define a rich collection of arrival sequences.

These approaches are intended for multiprocessor architectures. Therefore they do not consider network operations such as data transfers between resources. The analytical methods focus on predicting maximal/minimal bounds of the systems whereas we want to analyze also the expected (most probable) behavior of the applications. The approaches do not provide automatic support for features typical for distributed systems, e.g., task replication. In addition, each task is supposed to emit a new data item for each processed data item. This is a limiting factor since we work with event detection tasks which emit new data only if an event is detected.

In [2], an extensive survey of models and algorithms for workflow scheduling is given. They organize various characteristics of the applications into three components: the workflow model, the system model, and the performance model.

Another survey of workflow models is carried out in [11]. They propose several novel taxonomies of the workflow scheduling problem, considering five facets: workflow model, scheduling criteria, scheduling process, resource model, and task model.

Both surveys provide a general structured look on the known results in the area of workflow modelling and scheduling. They do not mention any approaches to handle variable processing costs of a single task nor variable data sizes output by a single task. Also, the tasks are assumed to emit a new data item for each processed data item. In our research, we focus on a specific area of workflow systems (event detection in multimedia streams) whose characteristics are not completely dealt with by the current approaches.

Modelling and evaluation of processing strategies is considered in the papers mentioned above too. In [3], a processing strategy based on the fixed task priority and length of queues is studied. Event count automata [4] allow to specify a variety of processing strategies, however, a systematic way is not provided. Periodic schedules [2] provide another way to specify processing strategies. Another type of specification consists of giving the fraction of the time each processor spends executing each task [2]. The two previous types of definition do not consider dynamic features of the applications (e.g., variable queue lengths). In our model, it is possible to define various processing strategies in a formal way while taking into account also the dynamic properties of running applications.

All the related work above is specifically targeted at streaming applications. Representatives of a general approach are Petri nets and their extensions which have been widely used as tools for analysis of distributed systems. In [10], Queue-ing Petri Nets are used to estimate throughput in computer networks of modern data centers. Colored Petri Nets are utilised for analysis of workflow models in [4,6].

In summary, there are both general and specific models and methods for performance analysis of stream processing systems. However, we see a lack of approaches depicting specific features of distributed streaming applications which deal with multimedia data (e.g., variable processing costs).

2   Model

In this section, we present a model of stream processing applications adopted to systems working with multimedia data. More details about the model can be found in our previous work [9]. The model is focused on those aspects of tasks (atomic computation units), data, and underlying network of resources which enable us to derive performance qualities of the applications.

The whole model consists of three different perspectives of the applications. The workflow model describes streams of data and tasks which process the streams. The system model is used to characterize the infrastructure of the network of available computational resources. The deployment model puts both models together and represents a mapping of individual tasks to computational resources. In addition, the deployment model is used to express processing strate-gies of the tasks at the resources.

Here is a list of the model features:

- Workflow model
    - Workflow graph (task dependencies)
    - Processing cost (the cost to process data items)
    - Output frequency (how frequently new data items are output)
    - Data size (the size of generated data items)
- System model
    - Topology of the underlying network (resource connectivity)
    - Resources (computational power)
    - Connections (bandwidth)
- Deployment model
    - Deployment graph (task to resource mapping)
    - Processing strategy (task/stream prioritization)

2.1   Workflow Model

The workflow model is based on a directed acyclic graph where nodes repre-sent tasks, and edges show flow of data streams between the tasks. Nodes with no income edges represent entry points of the streams to the application. An example of such a graph is in Fig. 1a.

We consider the following aspects of the tasks and the streams which are cru-cial for performance analysis. We need to know the processing costs of individual data items at each task, i.e., how much computational resources are required. Other important aspects are sizes of data items transferred between the tasks and also how frequently the data are transferred (i.e., how often the tasks output data to their successors).

To capture the variability in processing costs, we define the highest bound of processing cost cumulatively for individual lengths of data sequences. Formally, $cost(\Delta) = x$ where x is the highest possible number of processing units (e.g., CPU cycles) which are needed to process any sequence of data items of length Δ. Analogically, the cost function can be defined to set the minimal limits. In addition to the maximal and minimal limits defined above, also the probability density cost function is specified so that it is possible to work with the distribution of possible costs during performance analysis.

Similarly, we define the cumulative limits and distribution of data sizes which are transferred between two given tasks. The output frequency is also specified by cumulative limits and probabilities of the number of output data items. The frequency can be based on the number of processed data items, or it can be time dependent.

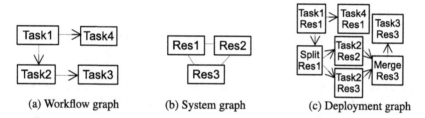

(a) Workflow graph   (b) System graph   (c) Deployment graph

Fig. 1. Graph examples of workflow and system models, and a corresponding deployment model

2.2   System Model

The system model represents the infrastructure of the network of available computational resources. Only two components of this model are differentiated: computational resources and connections between them. The topology of the network is modelled by an undirected graph where nodes represent computational resources, and edges show their connections via the computer network (see Fig. 1b).

2.3   Deployment Model

Finally, we have to specify which task is placed on which resource. This can be represented as a directed acyclic graph. Each node is of the form (t, r) where t is a task and r is a resource where the task is placed. The edges represent flow of the data. Each task has to be placed on at least one resource. For the cases when a task is put on more than one resource, we define two special types of nodes: $(split, r)$ and $(merge, r)$.

Node $(split, r)$ splits one stream into two or more partial streams. The distribution of data items to the partial streams is specified by cumulative limits and probability density function. Node $(merge, r)$ merges two or more streams output by different instances of the same task. See an example of the deployment graph in Fig. 1c.

We extend the original model with processing strategies to describe how the performance is affected when multiple tasks are collocated on the same resource. The processing strategies are used, for instance, to decide whether and which of the tasks should be prioritized. Also, the processing strategy takes place when there are two or more data items to be processed by the same task. The question may be whether only one data item should be processed at a time or whether the data items can be processed in parallel. The processing strategy specifies how much computational power should be dedicated to processing individual data items at the resource. The assigned computational power is expressed in terms of processing units which should be subtracted from the remaining processing cost of a data item.

Let k be the number of tasks at a given resource. Suppose the tasks are numbered 1 to k. We define

$$
\begin{aligned}
ProcessingStrategy_{Resource}(((t_{11}, t_{12}, \ldots, t_{1i_1}), (t_{21}, \ldots, t_{2i_2}), \ldots, \\
(t_{k1}, \ldots, t_{ki_k}))) = ((c_{11}, c_{12}, \ldots, c_{1i_1}), (c_{21}, \ldots, c_{2i_2}), \ldots, (c_{k1}, \ldots, c_{ki_k}))
\end{aligned}
\tag{1}
$$

The *ProcessingStrategy* function returns the amount of processing units which are subtracted from remaining processing cost of individual data items at the resource per time unit. Each tuple in the input and output corresponds to a single task. Elements of the input tuple $(t_{j1}, t_{j2}, \ldots, t_{ji_j})$ represent data items being currently processed by the task j; the order of the data items is given by their arrival time, i.e., the earliest ones are in the beginning. Each t_{jm} is a pair $(remainingCost, elapsedTime)$ where $remainingCost$ denotes the remaining processing cost and $elapsedTime$ represents the time since the data item arrived at the task. Elements of the output tuple $(c_{j1}, c_{j2}, \ldots, c_{ji_j})$ symbolize the amount of processing units which are subtracted from remaining processing cost of corresponding data items of the task j at the resource per a time unit.

This definition allows to specify a wide range of processing strategies. The strategies can be based on the order in which the data items arrive (e.g., FIFO); the data items can be prioritized according to their remaining costs (e.g., least remaining cost); the elapsed time information can be used to prevent items from starvation.

To give a specific example, suppose there are two tasks A and B placed on a given resource. Each task processes its data items sequentially (i.e., one at a time). The task A has a priority over B, and consumes 70% of available computational power. Suppose, the resource can process up to 10 processing units per a time unit. Then

$$ProcessingStrategy_{Res}(((t_{A1}, t_{A2}, \ldots, t_{Ai_A}), ())) = ((10, 0, 0, \ldots, 0), ()) \quad (2)$$

$$ProcessingStrategy_{Res}(((), (t_{B1}, t_{B2}, \ldots, t_{Bi_B}))) = ((), (10, 0, 0, \ldots, 0)) \quad (3)$$

$$ProcessingStrategy_{Res}(((t_{A1}, t_{A2}, \ldots, t_{Ai_A}), (t_{B1}, t_{B2}, \ldots, t_{Bi_B}))) = \\ ((7, 0, 0, \ldots, 0), (3, 0, 0, \ldots, 0)) \quad (4)$$

The first and second lines define the behavior when data items of just one stream are available; they express that 10 processing units of the first data item in the available stream are subtracted from the remaining processing cost. The third line describes the situation when data items of both streams are available. The first data item of the first stream consumes 7 processing units; the first data item of the second stream consumes 3 processing units.

A similar approach can be adopted to specify also data transfer through connections. This is useful to define, e.g., how many data items can be transferred concurrently, or if any stream should be prioritized.

3 Performance Analysis

As soon as an application is described using the presented model, we may proceed to the performance analysis. For this purpose, the model is converted to a Colored Petri Net (CPN) [7] which is an extension of the standard Petri nets.

Once the CPN is created, it can be analyzed using state space exploration techniques (e.g., [8]). If a technique for a full space exploration is used, the state space explosion problem is likely to emerge [4]. Therefore, we focus on simulation based techniques in our experiments for the sake of efficiency.

3.1 Colored Petri Nets

Colored Petri Nets [7] is a discrete-event modelling language combining Petri nets with the functional programming language Standard ML (SML). Petri nets provide basic primitives for modelling concurrency, communication, and synchronisation; Standard ML provides primitives for the definition of data types and describing data manipulation. CPNs allow to model a system as a set of hierarchically related modules; it also has a support for time representation in the modelled systems.

3.2 Model to CPN

In this part of the paper, we show how an application represented by the presented model can be described as a Colored Petri Net.

The final CPN is built gradually by creating small modules and joining them into bigger ones. The structure of the modules is depicted in Fig. 2a. Each module is a CPN which is able to communicate with its parent and children. A child module is represented by a transition (the so called substitution transition) in its parent. Each non-root module defines its input and output places. Tokens of these input and output places are simultaneously present also in the corresponding input and output places of the parent's substitution transition and vice versa.

The generic algorithm how the model is converted to the CPN is captured by Algorithm 1. The CPN is generated in a top-down manner. The *application* module is built at first; then a module for each substitution transition is generated recursively.

Algorithm 1. Model to CPN

---

function BUILDCPN(*model*) **return** *buildModule(null, model)*
function BUILDMODULE(*substitutionTransition, model*)
    module ← createModule(substitutionTransition, model)
    for all *trans ∈ module.getSubstitutionTransitions()* **do**
        module.addSubModule(buildModule(trans, model))
    return *module*

Let us explore the individual modules. Note that the provided figures of the modules are not precise models of CPNs, but they rather depict the schema of the modules for the sake of simplicity.

At the top of the hierarchy, there is a view of the whole application. On this level, it can be observed how the streams are sent through connections between resources (see Fig. 2b). This module is generated according to the deployment graph. The places are used to keep data items of the streams; the transitions represent resource and connection modules. Also splits and merges of the streams can be observed in this module which are defined in the deployment model.

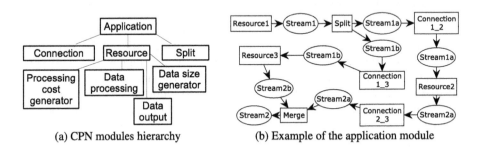

(a) CPN modules hierarchy   (b) Example of the application module

Fig. 2. Model to CPN

The *connection* module (see Fig. 3a) is used to simulate transfer of data items between two resources. The contents of the streams are passed by the *application* module. Each data item of the input streams carries its remaining data size to be transferred. *Transfer* decreases the remaining data size of individual data items in the input streams based on the policy defined in the deployment model (processing strategy). Once the remaining data size reaches zero, the data item is sent to the corresponding transferred stream whose content is reflected back to the *application* module. The *connection capacity* ensures that the maximum capacity of the connection specified in the system model is not exceeded per a time unit.

The *resource* modules are made of several smaller modules covering the properties defined for the workflow, system and deployment models. Figure 3b shows an example of the *resource* module. Contents of the input streams are passed by the *application* module. Once a data item enters the *resource* module, its processing cost is generated. Then, the data item enters the *process* module which simulates the actual processing according to the processing strategy as defined in the deployment model. After that, new data items are generated based on the output frequency policy specified in the workflow model. We can see that multiple output streams may be generated by a single *output* module (Output1). Finally, data sizes are assigned to the new data items. *Output3* in the figure represents a time based output generator which is especially useful for generating input streams of the system. It does not have any input streams since it uses just the notion of time to generate new items.

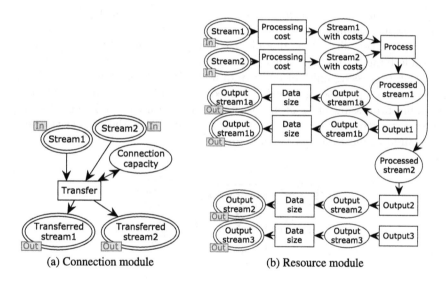

(a) Connection module   (b) Resource module

Fig. 3. Connection and resource modules

Figure 4a shows a schema of a *processing cost generator*. Its task is to generate a cost for each data item of the input stream which is subsequently used by the *data processing* module. The *upper* and *lower limits* keep the data of maximal and minimal cumulative functions respectively. *Cost probabilities* store data of the probability density cost function. *Cost history* contains costs assigned to recent data items. All these data serve as input to the underlying SML expressions which generate the final cost and update the *cost history*. In the current implementation, the cost is created randomly based on the probability density cost function, and it is validated whether it fits between the upper and lower limits. The limits and probabilities are based on the corresponding functions in the workflow model. *Data size generators*, *data output* and *split* modules are modelled analogically. Figure 4b shows a module for generating new data items based on time. The generated *time* is passed to *timer* which ensures that the *generate data item* transition is enabled after *time* elapses.

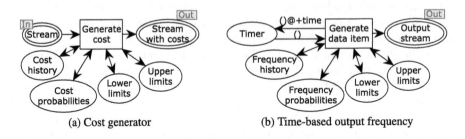

(a) Cost generator   (b) Time-based output frequency

Fig. 4. Cost generator and time-based output frequency

The schema of the *data processing* module is depicted in Fig. 5. Each data item of input streams keeps information about its remaining processing cost. The core functionality of the module is contained within the *process* transition. The underlying SML expressions of the *process* transition are used to decrease remaining processing units of input data items. The amount of decreased units is based on the processing strategy defined in the deployment model. Once the remaining processing units of a data item reach zero, the data item is sent to the processed stream. The *resource capacity* ensures that the *process* transition can be fired at most once per a time unit.

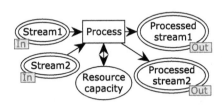

Fig. 5. Data process module

The final CPN captures all the aspects of the model presented in Sect. 2. As of now, the conversion of the model to CPN is done manually. In future, we intend to implement an automatic converter.

4   Experiments

To validate the proposed model and its representation in CPNs, two experiments were conducted. For each of them, a stream processing application was created and deployed in a distributed environment. During the run of the applications, performance related characteristics were gathered. Also the applications were modelled using the proposed approach and subsequently converted to a CPN. After that, simulations of the CPNs were run, and performance related properties were derived. Finally, the results were compared. The experiments are focused on the validation of a representation of the processing strategy since this is a new feature of the proposed model. Other experiments validating the precision of the model can be found in our previous work [9].

Both the applications consume a stream of images and detect faces in them. If any faces are found, feature descriptors of the image are extracted (Fig. 6a). The applications were implemented using the Apache Storm technology [1] and deployed as a Storm topology in a distributed environment. The Storm cluster is run on 2 virtual machines, each with 4 available CPUs and 4 GB RAM. The virtual machines are managed by OpenNebula[1] cloud manager (Fig. 6b).

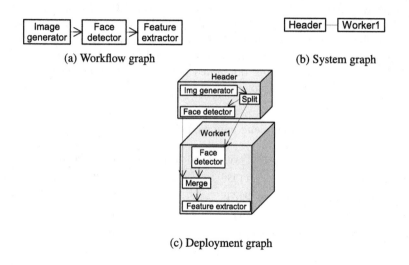

(a) Workflow graph

(b) System graph

(c) Deployment graph

Fig. 6. Experiment graphs

Figure 6c shows the deployment graph. The *image generator* is placed on *Header* where the dataset is present. The *face detector* is duplicated and placed on *Header* and *Worker1* so that the images can be processed in parallel. The split node distributes 70 % of the images to *Header* and 30 % to *Worker1*. The *feature extractor* receives merged faces stream on *Worker1*. To model the transfer of data items between the resources, the FIFO strategy is used, i.e., the data item with the lowest *elapsedTime* is sent. As there are two tasks placed on *Worker1*, a processing strategy should be specified. In the experiment #1, both the *face detector* and the *feature extractor* have the same priority. In the experiment #2, the *face detector* task is prioritized by adjusting the processes' priorities in the

---

[1] http://opennebula.org/.

operating system. The processing strategy when at least one data item of each stream is available at the resource is the following:

$$ProcessingStrategy_{Worker1}(((t_{11}, t_{12}, \ldots, t_{1i_1}), (t_{21}, t_{22}, \ldots, t_{2i_2}))) = \\ ((1, 0, 0, \ldots, 0), (0.4, 0, 0, \ldots, 0)) \tag{5}$$

where t_{1k} are data items belonging to the *face detector*, and t_{2k} items belong to the *feature extractor*.

During the run of the applications in Storm and also during simulations of the CPN, delays at the *feature extractor* were measured, i.e., the time since an image is output by the *image generator* until it is processed by the *feature extractor*.

In Table 1, we can see the results concerning the maximum (minimum) measured delays. Each row is labelled by either CPN or Storm; the columns correspond to the two experiments. In the CPN rows, there are the maximum (minimum) delays retrieved during the simulation of the CPN. In the Storm rows, there are the percentages of data items whose delay was equal or less (greater) than the delay predicted by the CPN. It can be observed that nearly 100 % of the data items processed by Storm were processed within the limits set by the CPN. We can also notice a big difference between maximum delays when the *face detector* is prioritized or not.

Table 1. Maximum and minimum delays at CPN and Storm

	Face det. not prioritized	Face det. prioritized
CPN: max delay [ms]	11586	22636
Storm: % of delays $\leq$ CPN max	99.70	99.97
CPN: min delay [ms]	1447	1536
Storm: % of delays $\geq$ CPN min	100	100

In Fig. 7, there are relative frequencies of the delays measured in Storm and retrieved during the CPN simulations. The values for the relative frequencies are computed as follows. The timeline is separated into time intervals (0.5 s long), and for each of the intervals, number of data items having the corresponding delay is counted. Finally, the number is divided by the overall amount of the data items to obtain the relative frequency. The frequencies are represented as a continuous line connecting the discrete values.

Figure 7a compares delay frequencies for both experiments measured in Storm; Fig. 7b depicts the frequencies measured in CPN. It can be observed that the most probable delays (5–6 s) were correctly predicted by the CPNs. When the *face detector* is not prioritized, this peak is more significant both in Storm and the CPNs. Also the frequency of delays between 2 and 3 s is greater when the *face detector* is not prioritized. In case of the *face detector* prioritization, we can notice higher frequencies of delays greater than 7 s both in Storm and the CPNs.

In summary, through CPN simulations, it was managed to predict maximal, minimal and most probable delays with high accuracy. According to the experiment results, the proposed model and its representation in CPNs can be used to compare different deployment scenarios of real streaming applications.

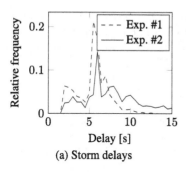

(a) Storm delays

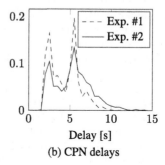

(b) CPN delays

Fig. 7. Experiment delays (Exp. #1 – no priority, Exp. #2 – face detector prioritized)

5   Conclusion

In this paper, we present performance analysis of distributed stream processing applications using Colored Petri Nets. We extended our previously proposed model with processing strategies which are used to specify performance effects when multiple tasks are placed on the same resource.

We show a detailed conversion of the whole proposed model to the CPNs. The accuracy of the model was validated through a couple of experiments dealing with multimedia data. CPN simulations results were compared to real measurements performed on applications running in a Storm cluster. Our model was able to predict nearly 100 % of the maximum and minimum delays precisely. Moreover, it was possible to reliably predict the distribution of the delays using the proposed methods.

Acknowledgements. This work was supported by the Czech national research project GBP103/12/G084. The hardware infrastructure was provided by the METACentrum under the programme LM 2010005.

References

1. Apache Storm. https://storm.apache.org/
2. Benoit, A., Çatalyürek, Ü.V., Robert, Y., Saule, E.: A survey of pipelined workflow scheduling: models and algorithms. ACM Comput. Surv. (CSUR) **45**(4), 50 (2013)
3. Bouillard, A., Phan, L.T., Chakraborty, S.: Lightweight modeling of complex state dependencies in stream processing systems. In: 15th IEEE Real-Time and Embedded Technology and Applications Symposium, RTAS 2009, pp. 195–204. IEEE (2009)

4. Chakraborty, S., Phan, L.T., Thiagarajan, P.: Event count automata: a state-based model for stream processing systems. In: 26th IEEE International Real-Time Systems Symposium, RTSS 2005, pp. 87–98. IEEE (2005)

5. Chakraborty, S., Thiele, L.: A new task model for streaming applications and its schedulability analysis. In: Proceedings of Design, Automation and Test in Europe, pp. 486–491. IEEE (2005)

6. Gottumukkala, R.N., Shepherd, M.D., Sun, T.: Validation and analysis of jdf work-flows using colored petri nets, 8 June 2010. US Patent 7,734,492

7. Jensen, K., Kristensen, L.M.: Coloured Petri Nets: Modelling and Validation of Concurrent Systems. Springer, Heidelberg (2009)

8. Kristensen, L.M., Petrucci, L.: An approach to distributed state space exploration for coloured petri nets. In: Cortadella, J., Reisig, W. (eds.) ICATPN 2004. LNCS, vol. 3099, pp. 474–483. Springer, Heidelberg (2004)

9. Nalepa, F., Batko, M., Zezula, P.: Model for performance analysis of distributed stream processing applications. In: Chen, Q., Hameurlain, A., Toumani, F., Wagner, R., Decker, H. (eds.) DEXA 2015. LNCS, vol. 9262, pp. 520–533. Springer, Heidelberg (2015)

10. Rygielski, P., Kounev, S.: Data center network throughput analysis using queueing petri nets. In: 2014 IEEE 34th International Conference on Distributed Computing Systems Workshops (ICDCSW), pp. 100–105. IEEE (2014)

11. Wieczorek, M., Hoheisel, A., Prodan, R.: Towards a general model of the multi-criteria workflow scheduling on the grid. Future Gener. Comput. Syst. **25**(3), 237–256 (2009)

GPU-Accelerated Real-Time Mesh Simplification Using Parallel Half Edge Collapses

Thomas Odaker[1]($\boxtimes$), Dieter Kranzlmueller[1], and Jens Volkert[2]

[1] Ludwig Maximilians Universitaet, Muenchen, Germany
odaker@a1.net, kranzlmueller@ifi.lmu.de
[2] Johannes Kepler University, Linz, Austria
jv@gup.jku.at

Abstract. Mesh simplification is often used to create an approximation of a model that requires less processing time. We present the results of our approach to simplification, the parallel half edge collapse. Based on the half edge collapse that replaces an edge with one of its endpoints, we have devised a simplification method that allows the execution of half edge collapses on multiple vertex pairs of a mesh in parallel, using a set of per-vertex boundaries to avoid topological inconsistencies or mesh foldovers. This approach enables us to remove up to several thousand vertices of a mesh in parallel, depending on the model and mesh topology. We have developed an implementation that allows to exploit the parallel capabilities of modern graphics processors, enabling us to compute a view-dependent simplification of triangle meshes in real-time.

1 Introduction

Highly detailed polygonal models are commonly used for visually appealing scenes. The triangle count still poses an important factor in performance considerations. A wide variety of simplification operators have been devised that can be used to reduce the complexity of 3-d models. Falling back on these operators a plethora of algorithms and approaches have been developed to create simplifications of given objects. Over the last decade various algorithms have been presented that are designed to utilize the parallel processing power of modern GPUs to speed up the simplification process (Papageorgiou and Platis [9]) with some focussing on calculating the simplified results in real-time such as Hu et al. [3] or DeCoro and Tatarchuk [4].

We present the results of our approach, the parallel half edge collapse. It is designed to provide a parallel solution to simplification of manifold triangle meshes that can be executed on a GPU.

1.1 Previous Work

Hoppe et al. [2] present the edge collapse. This operator replaces an edge of a triangle mesh with a single vertex, so effectively removing a vertex and one or two triangles. A more restrictive version is the half edge collapse, where the

© Springer International Publishing Switzerland 2016
J. Kofroň and T. Vojnar (Eds.): MEMICS 2015, LNCS 9548, pp. 107–118, 2016.
DOI: 10.1007/978-3-319-29817-7_10

position of the replacement vertex cannot be freely chosen, but is one of the endpoints of the collapsed edge. The edge collapse has the disadvantage that it may create foldovers in the mesh as well as topological inconsistencies [6].

In Hoppe [5], this operator and its inverse - the vertex split - are used to describe an algorithm to coarsen or refine a mesh. Given a detailed mesh Hoppe performs a series of edge collapses on it, storing the applied operations in a data structure. Then the mesh is represented using the coarse version and a series of refinement operations (vertex splits) that can be used to compute the desired refinement of the mesh. In Hu et al. [3] (with further explanation in [8]) this approach is adapted for execution on a GPU. While Hoppe defines a series of strictly iterative operations, Hu et al. replace Hoppe's data structure with a tree that defines dependencies between the precomputed edge collapses/vertex splits. This improved algorithm allows for a parallel execution and improved performance.

Another method of simplification is the cell collapse originally described in Rossignac and Borrell [1]. Here a number of cells is superimposed over the mesh with all vertices within a cell being combined into a single vertex. While this allows for fast generation of a coarse mesh, it has the disadvantage of ignoring the topology of the mesh which can result in low quality simplifications.

DeCoro and Tatarchuk [4] have adapted this approach to be executed on programmable graphics hardware. Their algorithm executes three passes over the mesh: cell creation, calculation of replacement position for each cell and generation of the decimated mesh.

Papageorgiou and Platis [9] present an algorithm that executes multiple edge collapses in parallel. However, they do not rely on precomputed data structures. Their approach divides a mesh into a series of independent areas. In each area an edge collapse can be safely executed without influencing another one. The algorithm is designed to be executed on a GPU, with the steps of computation of independent areas and performing a series of edge collapses in parallel being repeated until the desired simplification is achieved. This speeds up the computation, but does not provide real-time simplification.

In [10] we presented the concept of our approach, the parallel half edge collapse. In this paper we want to introduce further details of this algorithm and discuss the results of this approach in detail.

1.2 Algorithm Overview

The parallel half edge collapse aims to provide a view-dependent, real time simplification of a triangle mesh. The goal is to compute the simplification at run-time without relying on pre-computed operations. The simplification operator used is the half edge collapse. We determine a set of vertices R that are to be removed from the mesh and the complementary set S (remaining vertices of the mesh). Vertices are removed by performing half edge collapses on edges of the mesh that connect vertices in R with a vertex in S (removing the vertex in R and replacing the edge with the vertex in S), while executing as many of these operations as possible in parallel. To avoid mesh foldovers or topological inconsistencies - even when neighbouring vertices are removed at the same time - we compute a set of per-vertex boundaries to determine if a half edge collapse would cause any of the aforementioned issues. Since only vertices in R that have a neighbour in S can be removed using this approach, we apply multiple iterations of the parallel half edge collapse until all vertices in R have been processed. After each iteration, we execute a reclassification step. This analyses the remaining vertices in R taking changes in the topology and surface shape into account and may move them from R to S to avoid low quality simplifications.

2   Classification

The vertex classification analyses all vertices of a triangle mesh and assigns each one to either R or S. This step is divided into the vertex analysis and the initial classification.

Vertex analysis is performed before the simplification. For each vertex V an error value $e(V)$ is computed. The initial classification falls back on $e(V)$, applies scaling based on camera data like view vector and distance between camera and vertex and compares the result to a user-defined threshold u which leads to the creation of S and R.

2.1   Static Vertex Error

The static vertex error is calculated using a geometric error metric. We chose the average distance between the neighbouring vertices and the tangential plane of V for this purpose. The tangential plane is constructed from the position of V as well as the normal vector stored in V and expressed as $ax+by+cz+d = 0$ with $t = [a, b, c, d]$. For each neighbouring vertex N_i with the position $n = [n_x, n_y, n_z, 1]$ the quadratic distance from the tangential plane $d(V, N_i) = (t \bullet n)^2$ is calculated with $e(V)$ being the average quadratic distance.

$$e(V) = \frac{\sum_{i=1}^{m}(d(V, N_i))}{m} \qquad (1)$$

Since this metric is computed per-vertex and does not take a possible removal of neighbouring vertices into account, a large user threshold u could potentially select a large number of - if not all - vertices for removal. This could severely limit parallelism or in case of all vertices being marked for removal prevent a simplification at all. A metric manipulation is applied as a part of the error computation to avoid these issues.

2.2   Vertex Error Manipulation

This step aims to select a number of vertices from the mesh and assign them an error value $e(V) > u$ to guarantee their classification into S. We apply a

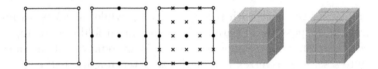

Fig. 1. Example point generation (left to right: layer L_0, layer L_1, layer L_2) and volumes for L_0 and L_1

layered version of the error manipulation. Layer L_0 contains the vertices that are assigned $e(V) > u$. Every additionally created layer selects additional points and manipulates their vertex error with a user selected value assigned to the layer.

For this approach a number of vertices has to be selected for each layer:

– Bounding box computation and creation of a set of points $P(L_0) = \{P_0^0, P_1^0, \ldots, P_n^0\}$ within the bounding box with equal axial distance $d(L_0)$ between points.
– Creation of additional layers L_i with points $P(L_i) = \{P_1^i, P_2^i, \ldots P_o^i\}$ created at halfway points between the points in L_{i-1} ($d(L_i) = \frac{d(L_{i-1})}{2}$, $i > 0$, 2-dimensional example on the left in Fig. 1).
– Generation of volume $B(P_j^i)$ for each point. $B(P_j^i)$ is centered around P_j^i and has a side length of $d(L_i)$ (trimmed to the bounding box). Right side of Fig. 1 shows the volumes for the first two sample layers.
– For each volume: determination of all vertices $V(P_j^i)$ within the volume
– For each volume: selection of one vertex from $V(P_j^i)$ and manipulation of the vertex error

Points P_j^i may not correspond to vertices of the mesh. For each P_j^i a vertex is selected from within the corresponding volume $B(P_j^i)$ if applicable. For a point P_j^i with position p_j^i and a vertex V_k with position v_k we calculate the weighted error $m(V_k, P_j^i)$ using the maximum side length l of the volume $B(P_j^i)$ as follows:

$$m(V_k, P_j^i) = (l - |p_j^i - v_k|)^2 * e(V_k) \tag{2}$$

For each volume the vertex with the maximum weighted error is used for error manipulation. The weighted error takes the static vertex error and the distance between P_j^i and V_k into account. The weighted error is designed to preferably select vertices closer to P_j^i, to achieve a more uniform distribution of vertices with a manipulated per-vertex error, while taking the original error value into account, to avoid keeping vertices with little influence to the surface shape in the simplified mesh.

3   Boundary Computation and Testing

The per-vertex boundaries are used to avoid half edge collapses causing mesh foldovers or topological inconsistencies. Due to the parallel nature of this approach, we cannot rely on simple methods of testing for such occurrences (e.g. a maximum rotation of triangle normals before and after a collapse).

3.1 Boundaries

The per-vertex boundaries $B(V)$ are a set of planes that is computed for each vertex V that has at least one neighbour in S (subsequently referred to as removal candidates). Each triangle containing V adds one or more planes to $B(V)$. Boundary planes are constructed based on how many vertices of a triangle are removal candidates (Boundary 1, 2 and 3 for 1, 2, and 3 removal candidates respectively) and use the camera position E. The parallel execution of half edge collapses has to be taken into account and the need for communication avoided. The planes are defined only to allow replacement positions for a removal candidate that would not cause a foldover or inconsistency, no matter what collapse - if any at all - is chosen for the other removal candidate in this triangle. This can lead to possible valid combinations of half edge collapses being blocked with boundaries 2 and 3.

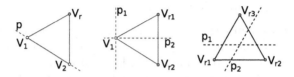

Fig. 2. Boundaries 1, 2 and 3

Boundary 1 (Vertices V_1, V_2 and removal candidate V_r, Fig. 2 left). A single plane p is constructed. It contains the vectors $\overrightarrow{V_1 - E}$ and $\overrightarrow{V_2 - E}$ as well as the points V_1 and V_2.

Boundary 2 (Vertex V_1 and removal candidates V_{r1}, V_{r2}, Fig. 2 middle). Two planes p_1 and p_2 are constructed. Plane p_1 is constructed using the vectors $\overrightarrow{V_{r1} - V_{r2}}$ and $\overrightarrow{V_1 - E}$ as well as the point V_1. Plane p_2 is constructed from $\overrightarrow{V_1 - E}$, $\overrightarrow{\frac{V_{r1}+V_{r2}}{2} - V_1}$ and V_1.

Boundary 3 (Removal candidates V_{r1}, V_{r2} and V_{r3}, Fig. 2 right). Two planes p_1 and p_2 are constructed for each removal candidate. They both contain the centroid S. For V_{r1}, plane p_1 contains the vectors $\overrightarrow{V_{r2} - V_{r1}}$ and $\overrightarrow{S - E}$ as well as the point S. Plane p_2 for V_{r1} is constructed from $\overrightarrow{V_{r3} - V_{r1}}$ and $\overrightarrow{S - E}$ and lies through the point S.

Each possible half edge collapse for a removal candidate V needs to be tested against these boundaries to avoid foldovers or topological inconsistencies:

- Selection of all triangles $T(V)$ containing V
- For each triangle in $T(V)$ determination of the appropriate boundary
- Construction of boundary planes and adding them to $B(V)$
- Testing of each possible half edge collapse against all planes in $B(V)$

Each possible half edge collapse for V has to be tested against each plane in $B(V)$ individually. If any intersection between the edge and any plane in $B(V)$ exists, the half edge collapse is considered invalid. This test can be simplified by adapting the orientation of the plane normals with regards to V. The test checks, if the dot product between the plane normal and the removal candidate V, as well as the dot product between the second vertex of an edge V' and the plane normal share the same sign. Planes are constructed so that the dot product of the plane normal and V have the same sign for all planes in $B(V)$. This reduces the test of a half edge collapse against a single plane in $B(V)$ to a dot product and the checking of the sign of the resulting value.

3.2 Half Edge Collapse Selection

The selection of one half edge collapse for each removal candidate is based on the approach by Garland and Heckbert [7]. We take Garland's and Heckbert's approach, compute their error value $\triangle(V')$ for each valid replacement position and execute the half edge collapse with the lowest error. While Garland and Heckbert update their vertex error by computing a new error value from the errors of V_1 and V_2, we deviate from this approach. We do not update the error value, but rather compute it at every step for the current intermediate mesh.

4 Deadlock Prevention and Reclassification

Since boundary 2 and 3 can block valid combinations of half edge collapses, it is possible for a deadlock to appear where two or more removal candidates mutually block each other and the simplification cannot be completed. Only boundary 1 always computes a correct result that does not block any valid half edge collapses. To avoid this, we create two sets of boundaries per removal candidate. $B(V)$ contains the boundaries as described above. $B'(V)$ is created with the planes of boundary 1 for all triangles containing V. This creates two results for each vertex: Result r_1 from checking with planes in $B(V)$ which allows to select a valid half edge collapse. Result r_2 from $B'(V)$ tells if any half edge collapse is allowed when not taking parallel execution into account. If r_1 and r_2 block all half edge collapses, the vertex cannot be removed and is reclassified. If only r_1 blocks all half edge collapses, the parallel execution prevents removal, the vertex remains in R and is considered as "ignored", excluding it from the removal candidates until at least one neighbour has been processed (either removed or reclassified).

After each removal step the list of removal candidates is updated. The initial vertex error is recomputed for all removal candidates using the current intermediate mesh. Since some neighbours of removal candidates are to be removed we adapt the metric here. We use the maximum distance between the tangential plane and the neighbours that are to remain in the mesh. Like the static vertex error, the updated vertex error is compared to the threshold and the vertex is reclassified if necessary.

5   Implementation

We have devised an implementation of the parallel half edge collapse using CUDA to be able to analyse our algorithm.

We had to extend per-vertex data for our algorithm. Each vertex stores a vertex error that is used to classify the vertex and updated for removal candidates after each iteration. Since boundary computation relies on the number of removal candidates in a triangle, it is also necessary to store if a vertex is "marked" as removal candidate.

The parallel half edge collapse requires information about all neighbouring vertices. It is necessary to create an additional buffer that serves as data storage containing a triangle strip for each vertex of the original mesh. The neighbouring triangles are used for two purposes during the execution of the parallel half edge collapse: boundary computation and calculation of the vertex error. Both applications need the geometric data, but do not rely on knowledge of the orientation of the triangle. Since the triangle fan is stored per vertex in this data structure, we can minimize storage requirements by only storing a list of neighbouring vertices in the correct order.

A second data structure is used to maintain a list of removal candidates. After the initial classification all removal candidates are determined by searching the neighbours of all those vertices, that are selected to remain in the mesh, for vertices to be removed. At the start of the removal step, the stored vertices are distributed among the threads executing the simplification. After the removal step has been completed, the list is repopulated with the new removal candidates for the next step.

We actually maintain two lists of removal candidates. Given that we assign each CUDA thread a vertex from this list, we can never guarantee that we actually have less vertices than cores. After the cores have completed processing the first assigned candidates, the list may still be partially populated. To avoid this issue, we use a separate input and output list, swapping them after each step.

6   Results

Figure 3 shows an example of a simplification of the Stanford Bunny in comparison to the original mesh (left) while Fig. 4 shows examples for other models that were simplified.

We calculated several simplifications of the Stanford Bunny to achieve comparable results, assess processing time and find bottlenecks and weaknesses. All measurements were taken using a Geforce GTX 670 GPU with 1344 cores. The original mesh of the Stanford Bunny consists of 35 947 vertices that form 69 451

Fig. 3. Original model (left), simplified version (about 93 % reduction in triangles, right)

Fig. 4. Simplifications of the model Armadillo, Dragon and Happy Buddha (93 %–95 % of triangles removed) from the Stanford 3-d scanning repository

Fig. 5. Wireframe models of test case 1–5. See Table 1 for details.

triangles. For the purpose of our measurements we compared 5 separate simplifications, ranging from 48 831 to 7 059 triangles. Beside the overall runtime and the number of triangles of the simplified mesh, we also analysed the number of iterations, including the number of vertices processed in each iteration. Given that our approach is executed on a GPU, we want a high number of processed vertices with each iteration to be able to fully utilize the cores of the GPU and increase the efficiency of the simplification. We measured the runtime of the individual steps of the simplification process and analysed the impact of the manipulation of the initial vertex error to the process.

Figure 5 shows the resulting wireframes for all 5 simplifications. Table 1 shows an overview of the results, including the number of triangles the simplified mesh is made up of, the number of triangles removed, the required iterations and the processing time in milliseconds. These results show the rise in necessary

Table 1. Test cases data overview (triangles, triangles removed, number of iterations and processing time in milliseconds)

	Triangles	Triangles rem.	Iter.	Time
1	48 831	20 620	2	1.94
2	41 732	27 719	3	2.13
3	29 014	40 437	5	3.36
4	17 565	51 886	8	4.43
5	7 059	62 392	12	5.76

Table 2. Data of simplifications of additional models (triangles, triangles removed and processing time in milliseconds)

Model	Triangles	Triangles rem.	Time
Bunny	69 451	62 392	5.76
Armadillo	345 944	323 356	29.1
Dragon	871 414	826 109	80.1
Buddha	1 087 716	1 022 232	96.1

 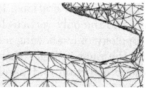

Fig. 6. Silhouette comparison

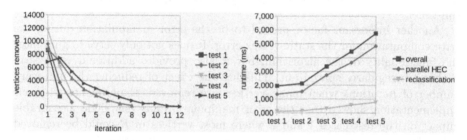

Fig. 7. Vertices removed in each iteration (left) and runtime analysis (right)

iterations to process all vertices marked by the vertex analysis. Table 2 shows the number of removed triangles and processing times for the additional models.

Figure 6 shows a comparison of a section of the image between the original (left) and a simplified version (right). This visualizes that the silhouette of the object is well preserved while the triangle count of the surface greatly reduced.

We further analysed the number of vertices processed in each step, which uncovered a problem with the execution of the parallel half edge collapse. A high grade of simplification causes a larger number of necessary iterations. We observed a high number of processed vertices in the early iterations of each simplification. During later iterations the number of vertices available for a half edge collapse dropped significantly.

The chart on the left in Fig. 7 shows the number of vertices that were removed in each step. As mentioned earlier, the GPU used for the test cases offers 1344 cores. The implementation assigns each CUDA thread an individual vertex to process and to remove. As this diagram illustrates, there are one or more steps in several test cases where not all cores of the GPU can be utilized due to an insufficient number of removal candidates. Especially test cases with a high number of triangles removed suffer from this problem. This issue may be further escalated by the mesh topology. A disadvantageous mesh topology can cause some vertices not to have a neighbour in S until most vertices marked for removal have been processed, effectively delaying the completion of the simplification process. Another issue we identified, that can cause this behaviour, is the deadlock prevention we implemented. As our approach can only recognize a possible deadlock once the subsequent iteration has started, the deadlock prevention could potentially delay the completion of the simplification process. Since it can mark a number of vertices as "to be ignored", the removal of those vertices may be distributed over several iterations, that might otherwise be unnecessary. In a worst-case scenario the only vertices left waiting for removal could be ignored ones with the topology only allowing a single removal per iteration.

The chart on the right in Fig. 7 shows the cumulative runtime of the reclassification, deadlock detection and parallel half edge collapse steps of the simplification process for all five test cases. It is obvious that the majority of the processing time is used for the execution of the parallel half edge collapse itself, while reclassification and deadlock prevention take up less than 20 % of the total run-time.

Another important factor proved to be the error manipulation during the error computation for the static vertex error. It does not only serve to guarantee the functionality of the algorithm, but it also provides additional vertices in S that are regularly spaced out. This has the effect of reducing the necessary number of iterations when many vertices are removed. Experiments with our implementation showed that the error manipulation has very little measurable impact on the test cases 1 and 2 where most vertices in R could be removed in the first iteration. In test case 5, however, the error manipulation caused a large reduction in necessary iterations, reducing them by a factor of 4, increasing parallelism and reducing run-time.

The last factor we analysed is memory usage. The parallel half edge collapse mainly relies on the fan data buffer as well as the buffers for the removal candidates and the vertex error that is stored in the vertex data. The execution needed 2.3 (Bunny), 11.2 (Armadillo), 29.9 (Dragon) and 35.2 (Buddha) megabytes for the tested models.

6.1   Comparison to Existing Algorithms

Given that the parallel half edge collapse is designed for computing the simplification in real-time, the most similar approaches are Hu et al. [3,8] and DeCoro and Tatarchuk [4]. While Papageorgiou and Platis [9] present an algorithm that aims to execute multiple edge collapses in parallel, they do not aim at real-time execution of the simplification. Even though their algorithm is faster than iterative approaches like the quadric error metric by Garland and Heckbert [7], they still take up to several seconds to compute the simplification. This fact leads to its omission for this comparison.

While Hu et al. provide real-time refinement of a triangle mesh, their approach is a parallel version of progressive meshes presented by Hoppe [5] and is not designed to calculate the complete simplification during rendering. It rather executes incremental changes in the form of vertex splits and edge collapses between frames. They report update times ranging from 10 ms (about 100 000 triangles) to nearly 70 ms (about 450 000 triangles) using a Nvidia GeForce 8800GTX GPU. Given that Hu et al. like Hoppe base their approach on a pre-simplified mesh that can be refined, their approach is to be described as bottom-up. The parallel half edge collapse on the other hand is a top-down approach. Due to these facts, a direct comparison of run times between these algorithms is not really meaningful.

DeCoro and Tatarchuk recalculate the simplification during image generation, but their approach is mainly designed to provide fast simplification time. It does not preserve manifold connectivity and tends to produce overall low quality (Fig. 8). Given that the parallel half edge collapse is a top-down approach, it has the disadvantage of higher execution time when producing coarser meshes. The approach by DeCoro and Tatarchuk has the advantage of producing a much more stable and predictable runtime. They report taking 13 ms to create a simplification of the Stanford Bunny on a "DirectX 10 GPU".

Fig. 8. Comparison with DeCoro and Tatarchuk [4] (right)

7   Future Work

The current calculation of the vertex error and its application during the initial classification only rely on geometrical data of the vertex. The classification of neighbouring vertices is not taken into account. As a result a large number of vertices can be marked for removal which can later be reconsidered during the reclassification phase.

As our analysis has shown, one of the major bottlenecks of our approach is the lack of removal candidates. Improving the initial classification to reduce the reliance on the reclassification step and providing a better set of removal candidates can be used to increase parallelism. Vertices that are reclassified during the execution in the current algorithm may provide additional removal candidates at the start of the simplification when an improved classification scheme is applied.

Another factor limiting the parallelism is the restriction of only executing half edge collapses between vertices in R and S. Allowing vertices with no neighbours in S to be subjected to a half edge collapse could be used to reduce the number of necessary iterations and therefore increase the parallelism of the approach.

8   Conclusion

The parallel half edge collapse has proven to allow fast, view-dependant simplification that can make use on the parallel processing power of modern GPUs by relying on isolated per-vertex operations. While our analysis has shown good results in terms of overall quality and execution time, it has also uncovered some limiting factors, namely the reduced parallelism that may be caused by the topology or a lack of removal candidates. Another limiting factor of the parallel half edge collapse lies in the execution of the simplification operator. Given that a vertex is chosen for removal and one half edge collapse selected for each removal candidate, it is not possible to select an "optimal" edge that is collapsed. While this causes iterative approaches to achieve a better overall quality, it is considered as a trade-off for the parallel execution and the performance gain of the parallel half edge collapse.

References

1. Rossignac, J., Borrell, P.: Multi-resolution 3D approximations for rendering complex scenes. In: Falcidieno, B., Kunii, T.L. (eds.) Modeling of Computer Graphics: Methods and Applications, pp. 455–465. Springer, Berlin (1992)
2. Hoppe, H., DeRose, T., Duchamp, T., McDonald, J.A., Stuetzle, W.: Mesh optimization. In: ACM SIGGRAPH Proceedings, pp. 19–26 (1993)
3. Hu, L., Sander, P.V., Hoppe, H.: Parallel view-dependent refinemnet of progressive meshes. In: Proceedings of the Symposium on Interactive 3D Graphics and Games, pp. 169–176 (2009)
4. DeCoro, C., Tatarchuk, N.: Real-time mesh simplification using the GPU. In: Proceedings of the Symposium on Interactive 3D Graphics, vol. 2007, pp. 161–166 (2007)
5. Hoppe, H.: Progressive meshes. In: ACM SIGGRApPH Proceedings, pp. 99–108 (1996)
6. Xia, J.C., El-Sana, J., Varshney, A.: Adaptive real-time level-of-detail-based rendering for polygonal models. IEEE Trans. Visual Comput. Graph. 3(2), 171–187 (1997)
7. Garland, M., Heckbert, P.S.: Surface simplification using quadric error metrics. In: SIGGRAPH Proceedings 1997, pp. 209–216 (1997)
8. Hu, L., Sander, P., Hoppe, H.: Parallel view-dependent level of detail control. IEEE Trans. Visual Comput. Graph. 16(5), 718–728 (2010)
9. Papageorgiou, A., Platis, N.: Triangular mesh simplification on the GPU. Vis. Comput. Int. J. Comput. Graph. 31(2), 235–244 (2015)
10. Odaker, T., Kranzlmueller, D., Volkert, J.: View-dependent simplification using parallel half edge collapses. In: WSCG Conference Proceedings, pp. 63–72 (2015)

Classifier Ensemble by Semi-supervised Learning: Local Aggregation Methodology

Sajad Saydali[1](✉), Hamid Parvin[2], and Ali A. Safaei[1]

[1] Department of Computer Engineering, Qeshm Branch,
Islamic Azad University, Qeshm, Iran
saydali@qeshmiau.ac.ir
[2] Department of Computer Engineering, Mamasani Branch,
Islamic Azad University, Mamasani, Iran
parvin@iust.ac.ir

Abstract. A novel approach for automatic mine detection using SONAR data is proposed in this paper relying on a probabilistic based fusion method to classify SONAR instances as mine or mine-like object. The proposed semi-supervised algorithm minimizes some target functions, which fuse context identification, multi-algorithm fusion criteria and a semi-supervised learning term. Our optimization purpose is to learn contexts as compact clusters in subspaces of the high-dimensional feature space through probabilistic feature discrimination and semi-supervised learning. The semi-supervised clustering component appoints degree of typicality to each data sample in order to identify and reduce the influence of noise points and outliers. Then, the approach yields optimal fusion parameters for each context. The experiments on synthetic datasets and standard SONAR dataset illustrate that our semi-supervised local fusion outperforms individual classifiers and unsupervised local fusion.

Keywords: Supervised learning · Ensemble learning · Classifier fusion

1 Introduction

Several wars and military fights over the last century have occurred around one hundred million unexploded land mines in approximately seventy different countries [1]. Furthermore, approximately five million new mines are yearly buried in the ground [3]. Most of these mines are anti-personal, and claim the lives of around 70 civilians on daily basis, in regions like Afghanistan, Angola, Bosnia, and Cambodia [1]. The main reasons behind these land mines proliferation include their tactical and psychological effectiveness, and their simple and low cost fabrication. In other words, they emerged as an interesting alternative for country and armed organizations which cannot acquire sophisticated defense systems [4]. Hence, in spite of mine-clearing attempts around the world, million landmines are still deployed. According to the United Nations, over 1000 years would be required to clear out the mine fields around the world by conventional mine neutralization methods [3]. The classical detection methods of buried land mines i.e. hand probes and metal detectors cannot be utilized for large operations. In addition, recent anti-personal mines are small sized and plastic made with small

© Springer International Publishing Switzerland 2016
J. Kofroň and T. Vojnar (Eds.): MEMICS 2015, LNCS 9548, pp. 119–132, 2016.
DOI: 10.1007/978-3-319-29817-7_11

metal portion making their detection challenge even more critical [3]. Other techniques like ground-penetrating radar and dogs trained to sniff out explosives turned out to be slow and dangerous because they should operate very close to mines to be able to detect them. Through the past decade, infrared camera based detection of buried mines proved to be efficient when the field conditions are suitable. Moreover, novel mine neutralization technologies include infrared emission, thermal neutron activation, energetic photon detection, and ground-penetrating radar [3]. The most recent studies seek to consolidate these technologies and conventional metal detectors in order to improve their detection performance [6].

For submarine naval mines, the aquatic mammals such as Atlantic and Pacific bottlenose dolphins, white whales, and sea lions have been used by the United States navy to discover buried mines [5]. On the other hand, shallow water of the surf zone impacts the performance of these mammals dramatically. Nowadays, a sound propagation technology called the SOund NAvigation and Ranging (SONAR), has been used for underwater mine detection. Considerable researches related to the detection and classification of SONAR signals has been done increasingly [2]. Side-scan sonar photography transfer high resolution shots of the sea floor scene. Although the these images have contributed to applications promotion like Mine Counter Measure (MCM) where speed is a key factor, they have not facilitated that much the detection of objects of interest in these scenes. In factuality, due to the acoustical medium and the wide heterogeneity of this environment for underwater scenes, the process of object detection and classification is even more demanding. Furthermore it is difficult to detect the objects located on the sea floor or buried under the sand inasmuch as their appearance may vary considerably based on the nearby scene. In spite of these challenges, SONAR imagery would operate in poor visibility conditions without additional light equipment, and then make researchers endeavor to develop submarine mine detection systems based on this image modality. Moreover, the low fabrication cost along with the low power consumption represents other main advantages that encouraged researchers to include SONAR module in Autonomous Underwater Vehicles (AUVs) for submarine mine detection. Processing this high-dimensional data on board has become an urgent need for this solution. Also, embedding unsupervised decision-making features and reducing the expert involvement in the recognition process emerged as new active research field. Novel clearing operations of underwater mines would rely on AUVs equipped with SONAR capabilities, and adopting Computer Aided detection (CAD) solutions. The recognition task consists in classifying signatures of region of interest as mine or not. The subsequent classification algorithms are intended to recognize the false alarms as possible while detecting real mines. Choosing an appropriate supervised classification model for sonar data pattern recognition is a critical issue for objects of interest detection under the sea [7].

In this paper, a probabilistic based local approach is presented that adapts multi-classifier fusion to different regions of the feature space. The proposed approach, first categorizes the training samples into different clusters based on the subset of features used by the single classifiers, and their confidences. This phase is a complicated optimization problem which is prone to local minima. We utilized a semi-supervised learning term in the proposed target function to relieve this problem. The process of categorization associates a probabilistic membership, representing the degree of typicality, with each data sample in order to recognize and decrease the impact of noise points and outliers [8]. Additionally, an expert is assigned to each obtained cluster. These experts represent the best classifiers for the corresponding cluster/context. Then, aggregation weights are estimated by the fusion component for each classifier. These weights represent the performance of the classifiers within all contexts. Finally, for a given test sample, the fusion of the individual confidence values is obtained using the aggregation weights associated with the closest context.

2   Proposed Method

Today, the notion of classifier fusion has opened promising windows, and has outperformed single classifier systems both theoretically and experimentally. The characteristic feature of a classifier ensemble is the classifiers evolutionary which makes them to outperform individual learners by inheriting their strengths and restricting their weaknesses. However, diversity and accuracy are two critical conditions for fusion algorithms to outperform individual classifiers. Classifier fusion, which relies on appropriate aggregation of classifiers outputs, treats all learners equally trained and competitive over the feature space. For testing, single expert classifications are yielded simultaneously, and the resulting outputs are aggregated to perform a final fusion decision. Typical fusion approaches include Borda count, average, majority vote, Bayesian, probabilistic, and weighted average.

Different approaches of combining learners can be divided into two major classes: local approaches and global approaches. Assuming that higher classification accuracy can be attained using appropriate data-dependent weights, local approaches consider a relevance degree to the different training set subspaces, but global approaches assign an average relevance degree for each learner with respect to the whole training set. Local fusion approach is required to put the input data samples into homogeneous clusters during the training phase. This way of clustering can be achieved on the space of individual learner classification, based on which classifiers behave similarly, or using attributes of the input space. Then, the most accurate classifier is to be as an expert in each space region. For classification, unknown instances are assigned to regions, and the final decision is made by the corresponding expert learner of this region. A dynamic data partitioning and classification during the testing phase is proposed here. The authors estimate the classifier accuracies using sample nearby the local regions of the feature space, and the most accurate one is used to predict the class of the test sample. The local fusion approach called Context-Dependent Fusion (CDF) starts by clustering the training instances into homogeneous categories of contexts. This clustering phase and selecting a local expert learner are two independent stages of CDF, respectively. The researchers have described a generic framework for context-dependent fusion which simultaneously clusters the feature space, and aggregates the outputs of the expert learners. This combining approach uses a simple linear aggregation to generate fusion weights for individual learners. Nonetheless, these weights may be ineffective to capture the classifiers mutual interaction.

An approach is outlined that evaluates the performance of each expert performance in local regions of the feature space. The most accurate classifier is acquired to predict the final decision for each local region. However, the performance evaluation for test instances is timely complex, and impacts the performance of the approach with large data. The clustering and selection phase determines the statically most accurate classifier. First, the decision regions are discovered by clustering the training instances. Next, the most accurate learner on this local region is opted. The main disadvantage of this solution is that it does not deal with more than one classifier per region. The researchers have extended the clustering and selection algorithm so it splits the training dataset into correctly and incorrectly categorized instances. The feature space is then grouped by clustering the training instances. For testing, the most effective classifier in the vicinity of the test sample is chosen in order to generate the final decision. In other words, each learner exploits its corresponding cluster. This approach decreases the computational effectiveness of the solution. Recently, in [11], the authors outlined a local fusion approach that divides the feature space into homogeneous clusters based on their features, and takes into account the acquired clusters while aggregating individual learners' outputs. The fusion phase includes appointing an aggregation weight to each individual learner corresponding to each context based on its relative performance. Note that the applied fuzzy approach increases the sensitivity of the fusion component to outliers which decrease the overall classification performance. Some authors proposed a probabilistic based local approach that adapts the fusion method to different regions of the feature space. It put the training samples into different clusters based on the subset of features used by the individual classifiers, and their confidences. This clustering phase generates probabilistic memberships representing the degree of typicality of each data sample in order to identify and discard noise points. Additionally, an expert classifier is associated with each obtained cluster. More specifically, each classifier learns to aggregate weights simultaneously. Finally, for a given test sample, the fusion of the individual confidence values is obtained by using the aggregation weights associated with the closest context/cluster. Although this approach had promising results, the adopted optimization approach is prone to local minima.

3   Local Fusion Based on Probabilistic Context Extraction

Given $T = \{t_j | j = 1, \ldots N\}$ as the desired output of N training observations. These outputs were obtained using K classifiers. Each classifier k uses its own feature set $X_k = \{x_j^k | j = 1 \ldots N\}$ and generates the confidence values $Y_k = \{y_j^k | j = 1 \ldots N\}$. The K feature sets are then concatenated to generate one global descriptor, $\chi = \bigcup_{k=1}^{K} \chi^k = \{x_j = [x_j^1, \ldots, x_j^K | j = 1, \ldots, N]\}$. The probabilistic-based context extraction for local fusion algorithm in [8] has been formulated as partitioning the data into C clusters minimizing one objective function. However, this clustering approach needs to estimate various parameters applying complicated optimization and is prone to several local minima. To overcome this potential drawback, we propose a semi-supervised version of the algorithm in [8]. The supervision information relies on two sets of pairwise constraints. The first one is *Should-link* constraints which specify that two data instances are expected to belong to the same cluster. The second set of constraint is the

ShouldNot-link which specifies that two data instances are expected to belong to different clusters.

Let *SL* be the set of *Should-link* pairs of instances. If the pair (X_i, X_j) belongs to *SL*, then X_i and X_j are expected to be assigned to the same cluster. Similarly, let *NL* be the set of *ShouldNot-link* pairs. If the pair (X_i, X_j) belongs to *NL*, then X_i and X_j are expected to be assigned to different clusters. In this work, we reformulate the problem of identifying the C components/clusters as a constrained optimization problem. More specifically, we modify the objective function in [8] as follows

$$
J = \sum_{j=1}^{N} \sum_{i=1}^{C} u_{ij}^{m} \sum_{k=1}^{K} v_{ik}^{q} d_{ijk}^{2}
$$

$$
+ \sum_{j=1}^{N} \sum_{i=1}^{C} \beta_i u_{ij}^{m} \left(\sum_{k=1}^{K} \omega_{ik} y_{kj} - t_3 \right)^2 + \sum_{i=1}^{C} \eta_i \sum_{j=1}^{N} (1 - u_{ij})^m \qquad (1)
$$

$$
+ \mu \left[\sum_{(X_t, X_k \in NL)} \sum_{i=1}^{C} u_{ij}^{m} u_{kj}^{m} + \sum_{(X_t, X_k \in SL)} \sum_{i=1}^{C} \sum_{p=1, p \neq i}^{C} u_{ij}^{m} u_{kj}^{m} \right]
$$

subject to $\sum_{i=1}^{C} u_{ij} = 1 \forall j, u_{ij} \in [0, 1] \forall i, j,$ $\sum_{k=1}^{K} v_{ik} = 1 \forall i, v_{ik} \in [0, 1] \forall i, k,$ and $\sum_{k=1}^{K} \omega_{ik} = 1 \forall i.$

In (1), u_{ji} represents the probabilistic membership of X_j in cluster i [8]. The $C \times N$ matrix, $U = [u_{ij}]$ is called a probabilistic partition if it satisfies:

$$
\begin{cases} u_{ij} \in [0, 1], \forall j \\ 0 < \sum_{i=1}^{C} u_{ij} < N, \forall i, j \end{cases} \qquad (2)
$$

On the other hand the $C \times d$ matrix of feature subset weight, $V = [v_{ik}]$ satisfies

$$
\begin{cases} v_{ik} \in [0, 1], \forall i, k \\ \sum_{k=1}^{K} v_{ij} = 1, \forall i \end{cases} \qquad (3)
$$

The first term in (1) corresponds to the objective function of the SCAD algorithm [8]. It aims to categorize the N points into C clusters centered in c_i such that each sample x_j is assigned to all clusters with fuzzy membership degrees. Also, it is intended to simultaneously optimize the feature relevance weights with respect to each cluster. SCAD term is minimized when a partition of C compact clusters with minimum sum of intra-cluster distances is discovered. The second term in (1) intends to learn cluster-dependent aggregation weights of the K algorithm outputs. ω_{ik} is the aggregation weight assigned to classifier k within cluster i. This term is minimized when the aggregated partial output values match the desired output. The third term in (1) yields the generation of the probabilistic memberships u_{ji} which represent the degree of typicality of each data point within every cluster, and reduce the effect of outliers on the learning process. In (1), $m \in (1, \infty)$ is called the fuzzier parameter, and values η_i are positive constants that control the importance of the third term with respect to the first and second ones. This term is minimized when u_{ij} are close to 1, thus, avoiding the trivial solution of the first term (where $u_{ij} = 0$). Note that $\sum_{i=1}^{C} u_{ij}$ is not constrained to sum to 1. In fact, points that are not representative of any cluster will have $\sum_{i=1}^{C} u_{ij}$ close to zero and will be considered as noise. This constraint relaxation overcomes the disadvantage of the constrained fuzzy membership approach which is the high sensitivity to noise and outliers. The parameter η_i is related to the resolution parameter in the potential function and the deterministic annealing approaches. It is also related to the idea of "scale" in robust statistics. In any case, the value of 0.7 determines the distance at which the membership becomes 0.5. The value of η_i determines the "zone of influence" of a point. A point X_j will have little influence on the estimates of the model parameters of a cluster if $\sum_{k=1}^{K} v_{ik}^2 \left(d_{ijk}^s \right)^2$ is large when compared with η_i. On the other hand, the "fuzzier" m determines the rate of decay of the membership value. When $m = 1$, the memberships are crisp. When $m \to \infty$, the membership function does not decay to zero at all. In this probabilistic approach, increasing values of m represent increased possibility of all points in the data set completely belonging to a given cluster. The last term in (1) represents the cost of violating the pairwise Should-link, and *ShouldNot-link* constraints. These penalty terms are weighted by the membership values of the instances that violate the constraints. Namely, typical instances of the cluster which have high memberships yield larger penalty term. The value of μ controls the importance of the supervision information compared to the other terms.

This algorithm performance relies on the value of β. Over estimating, it results in the domination of the multi-algorithm fusion criteria which yields non-compact clusters. Also, sub-estimating β decreases the impact of the multi-algorithm fusion criteria and increases the effect on the distances in the feature space. When appropriate β is chosen, the algorithm yields compact and homogeneous clusters and optimal aggregation weights for each algorithm within each cluster.

Minimizing J with respect to U is equivalent to minimizing the following individual objective functions with respect to each column of U:

$$J^{(i)}(U_i) = -\sum_{j=1}^{N} u_{ij}^m \sum_{k=1}^{K} v_{ik}^q d_{ijk}^2$$

$$+ \sum_{j=1}^{N} \beta_i u_{ij}^m \left(\sum_{k=1}^{K} \omega_{ik} y_{kj} - t_j \right)^2$$

$$+ \eta_i \sum_{j=1}^{N} (1 - u_{ij})^m \tag{4}$$

$$+ \mu \left[\sum_{(X_t, X_k \in NL)} u_{ij}^m u_{kj}^m + \sum_{(X_t, X_k \in SL)} \sum_{p=1, p \neq i}^{C} u_{ij}^m u_{kj}^m \right]$$

For $i = 1, \ldots, C$. By setting the gradient of $J^{(i)}$ with respect to the probabilistic memberships u_{ij} to zero, we obtain

$$\frac{\partial J^{(i)}(U_i)}{\partial u_{ij}} = m\left(u_{ij}\right)^{m-1} \left(\sum_{k=1}^{K} v_{ik}^q d_{ijk}^2 + \sum_{j=1}^{N} \beta_i \left(\sum_{k=1}^{K} \omega_{ik} y_{kj} - t_j \right)^2 \right.$$

$$\left. + \mu \left[\sum_{(X_i, X_k \in NL)} u_{kj}^m + \sum_{(X_i, X_k \in SL)} \sum_{p=1, p \neq i}^{C} u_{kp}^m \right] \right) \tag{5}$$

$$- m \eta_i \left(1 - u_{ij} \right)^{m-1} = 0 \tag{6}$$

This yields the following necessary condition to update u_{ij}:

$$u_{ij} = \left[1 - \left(\frac{D_{ij}^2}{\eta_i} \right)^{\frac{1}{m-1}} \right]^{-1} \tag{7}$$

where

$$D_{ij} = \sum_{k=1}^{K} v_{ik}^q d_{ijk}^2 + \beta_i \sum_{k=1}^{K} v_{ik}^q \left(\sum_{l=1}^{K} \omega_{il} y_{lj} - t_j \right)^2$$

$$+ \mu \left[\sum_{(X_t, X_k \in NL)} u_{kj}^m + \sum_{(X_t, X_k \in SL)} \sum_{p=1, p \neq i}^{C} u_{kp}^m \right]$$

D_{ij} represents the aggregate cost when considering point X_j in cluster i. As it can be seen, this cost depends on the distance between point X_j and the cluster's centroid c_i, the cost of violating the pairwise *Should-link*, and *ShouldNot-link* constraints (weighted by μ), and the deviation of the combined algorithms' decision from the desired output (weighted by β). More specifically, points to be assigned to the same cluster: (*i*) are close to each other in the feature space, and (*ii*) their confidence values could be combined linearly with the same coefficients to match the desired output.

Minimizing J with respect to the feature weights

$$v_{ik} = \sum_{t=1}^{K} \left[\left(D_{ik}^2 / D_{il} \right)^{\frac{1}{q-1}} \right] \tag{8}$$

where $D_{il} = \sum_{j=1}^{N} u_{ij}^m d_{ijl}^2$.

Minimization of J with respect to the prototype parameters, and the aggregation weights yields

$$c_{ik} = \frac{\sum_{j=1}^{N} u_{ij}^m X_{ik}}{\sum_{j=1}^{N} u_{ij}^m} \tag{9}$$

and

$$w_{ik} = \frac{\sum_{j=1}^{N} u_{ij}^m y_{kj} \left(t_j - \sum_{\substack{l=1 \\ l \neq k}}^{K} \omega_{il} y_{lj} \right) - \zeta_i}{\sum_{j=1}^{N} u_{ij}^m y_{kj}^2} \tag{10}$$

where ζ_i is a Lagrange multiplier that assures that the constraint in (2) is satisfied, and is defined as

$$\zeta_i = \frac{\sum_{k=1}^{K} \left(\sum_{j=1}^{N} u_{ij}^m y_{lj} \left(t_j - \sum_{k=1}^{K} \omega_{ik} y_{kj} \right) \right) \left(\sum_{j=1}^{N} u_{ij}^m y_{lj}^2 \right)^{-1}}{\sum_{l=1}^{K} \left(\sum_{j=1}^{N} u_{ij}^m y_{lj}^2 \right)^{-1}} \tag{11}$$

The acquired iterative algorithm starts with an initial partition and alternates between the update equations of u_{ij}; v_{ik}; w_{ik} and c_{ik} as shown in Algorithm 1.

Algorithm 1. The proposed semi-supervised probabilistic clustering, feature weighting and classifier aggregation.

> **Inputs**: *X: The data instances.*
>> *Y : The confidences obtained using the different classifiers.*
>> *NL: The set of ShouldNot-Link constraints.*
>> *SL: The set of Should-Link constraints.*
>> *T: The labels of the data instances.*
>> *C: The number of clusters.*
>> *m: The fuzzifier.*
>> *q: The exponent of the feature weights.*
>> *T: The labels of the data instances.*
>>> *β: The weight assigned to the second term of the objective*
>> *function (1).*
>>> *η: The weight assigned to the third term of objective function (1).*
> **Outputs**: *U: The probabilistic membership matrix of the data instances.*
>> *c_i: The Clusters centers.*
>> *V: The feature weights.*
>> *W: The aggregation weights.*

Begin
Initialize the centers;
Initialize the probabilistic partition matrix U;
Initialize the relevance weights;
Repeat
Compute d_{ijk}^2, for $1 \leq i \leq C$ and $1 \leq j \leq N$ and $1 \leq k \leq K$;
Update the relevance weights v_{ik} using equation (8);
Compute D_{ij}^2
Update the partition matrix U using equation (7);
Update the aggregation weights matrix W and the feature
Weights matrix V using equations (10) and (8), respectively;
Update the centers using equation (9);
Until (centers stabilize)
End

The time complexity of one iteration of this first component is $O(N \times d \times K \times C)$, where N is the number of data points, C is the number of clusters, d is the dimensionality of the feature space, and K is the number of feature subsets. The computational complexity of one iteration of other typical clustering algorithms (e.g. FCM, PCM) is $O(N \times d \times C)$. Since we use small number of feature subsets $(K = 3)$, one iteration of our algorithm has a comparable time complexity to other similar algorithms. However, we should note that since we optimize for more parameters, it may require a larger number of iterations to converge.

After training the algorithm described above, the proposed local fusion approach adopts the steps below in order to generate the final decision for test samples:

- Run different classifiers on the test sample within the corresponding feature subset space, and obtain decision values, $Y^j = \{y_{kj}|k = 1, \ldots k\}$.
- An unlabeled test sample inherits class label of its nearest training sample.
- Assign the membership degrees u_{ij} to the test sample j in each cluster i using Eq. (7).
- Aggregate the output of the different classifiers within each cluster using $\hat{y}_{ij} \sum_{k=1}^{K} w_{ik} y_{kj}$.
- The final decision confidence is estimated using $\hat{y} = \sum_{i=1}^{C} u_{ij} \hat{y}_{ij}$.

4 Experiments

We depicted the performance of the proposed semi-supervised local fusion algorithm applying synthetic data sets. Our approach is compared to individual classifiers and the proposed method in [8] for these datasets, in Table 1.

In this experiment, the need for semi-supervised probabilistic local fusion is illustrated. We utilize our semi-supervised local fusion approach to classify the synthetic 2-dimensional dataset. Let each sample be processed by two single algorithms (K-Nearest Neighbors (K-NN) with $K = 3$). Each algorithm, k, considers one feature X_k; and assigns one output value y_k. Samples from Class 0 are represented using blue dots and samples from Class 1 are displayed in red. Black samples represent noise samples. The dataset consists of four clusters. Each one of them is a set of instances from the two classes [9].

Table 1. Learned weights for each classifier with respect to the different clusters obtained using the method in [8] and the proposed semi-supervised method.

Method in [8]	Cluster number	1	2	3	4
	Classifier 1	31.15	73.43	2.89	2.31
	Classifier 2	68.88	26.89	97.33	97.74
Proposed method	Cluster number	1	2	3	4
	Classifier 1	29.49	73.63	2.93	2.41
	Classifier 2	69.99	26.49	97.23	97.61

To construct the set pairwise constraints, the samples being at the boundary of each cluster are selected randomly. 7 % of the overall numbers of instances are considered to be *Should-link* and *ShouldNot-link* sets. Pairs of instances belonging to the same cluster (based the ground truth) form the Should-link set. Similarly, pairs that belong to different clusters form the *ShouldNot-link* set.

Accuracy as performance measure is used to evaluate the performance of our semi-supervised method. The overall accuracy of the partition is computed as the average of the individual class rates weighted by the class cardinality. To take into consideration the sensitivity of the algorithm to the initial parameters, we ran the algorithm 10 times using different random initializations. Then, we computed the average accuracy values for each supervision rate. Based on experimentation, the accuracy increased at a much lower rate with supervision rate larger than 7 %. Hence, for the rest of the experiments we set the supervision rate used to guide our clustering algorithm to 7 %.

Table 2. Performance comparison of the individual learners, the method in [8], and the proposed method for SOANR data set [10].

	Accuracy	Precision	Recall	F-measure
KNN 1	82.69	76.80	88.03	82.03
KNN 2	84.16	79.72	88.42	83.84
KNN 3	85.11	81.42	88.08	84.61
Method in [8]	86.04	86.36	84.31	85.32
Proposed method	90.24	90.90	89.47	90.17

In this section, we use our approach to classify standard dataset frequently used by researchers from the machine learning community. Namely, we consider the SONAR dataset [10] which consists of 208 instances and 60 attributes. 97 instances were obtained by bouncing sonar signals off a metal cylinder under various conditions and at various angles. A variety of different aspect angles, spanning 90 degree for the cylinder and 180 degrees for the rock were considered to contain the dataset signal. Each attribute represents the energy within a particular frequency band, integrated over a given period of time. SOANR dataset is summarized in [10].

In our experiments, for individual learners and local fusion approaches we adopt a 5-fold cross-validation in which each fold is treated as a test set with the rest of the folds used for training.

We divide the SONAR features into three subsets, and we dedicate one learner for each one of them. We run simple K-NN learner to generate confidence values for each instance. We categorize the training samples using 3 K-NN classifiers (K = 3) within their corresponding feature subspaces. Then, the proposed semi-supervised local fusion is used to categorize the training instances into 3 homogeneous clusters, and learn the optimal aggregation weights. Then, test instances are classified using the three individual learners, and assigned to the closest cluster. Finally, the fusion decision is generated by combining the partial confidences with the aggregation weights of the closest cluster. Notice that Should-link and *ShouldNot-link* constraints are generated using a clustering algorithm. More specifically, we cluster the training dataset using the probabilistic-based algorithm in [12], and we include pairs of typical instances (with high probabilistic membership), belonging to the same cluster, in the *Should-link* set. On the other hand, pairs of typical instances (with high probabilistic membership), belonging to different clusters, are included in the *ShouldNot-link* set. We limit the number of pairwise constraints to 7 % of the total number of instances.

We report the mine detection accuracies, precision, recall and F-measure obtained using K-NN classifier with different values of the parameter K. As it can be concluded experimentally, $K = 5$ yields the best overall performance measures. Thus, for the rest of the experiments, we set this K to 5.

We compare the obtained average accuracy, precision, recall, and F-measure values obtained using individual K-NN learners, the method in [8], and the proposed method with the SONAR dataset in Table 2. Our semi-supervised approach outperforms the other classifiers on this dataset based on the four performance measures. This proves that the association of supervision information with local fusion technique yields better clustering results and let individual learners cooperate more efficiently to generate more accurate final decision. This confirms the results obtained with synthetic datasets in the previous experiment.

Table 3(a) shows the learners aggregation weights with respect to the various clusters generated by our algorithm. These weights reflect the impact of each individual learner within each cluster. For instance, the second individual K-NN is perceived by our approach as the most accurate classifier for instances from cluster 1. Similarly, the highest aggregation weight is assigned to the first individual K-NN within cluster 3.

Table 3. (a-above) Learned weights for each classifier in each cluster obtained using the proposed semi-supervised local fusion with SONAR data set [10]. (b-below) Per-cluster accuracy of the three K-NN classifers with SONAR data [10].

Cluster #	Cluster 1	Cluster 2	Cluster 3
K-NN 1	0.2080	0.2758	**0.8003**
K-NN 2	**0.6142**	**0.5309**	0.1038
K-NN 3	0.1778	0.1933	0.0959

Cluster #	Cluster 1	Cluster 2	Cluster 3
K-NN 1	0.6947	0.7523	**0.8846**
K-NN 2	**0.8713**	**0.8600**	0.5992
K-NN 3	0.6401	0.6765	0.5889

To indicate that the semi-supervised local fusion utilizes the strengths of the individual learners within local regions of the features space, we report the accuracy of the three individual learners (K-NN) within the 3 clusters. These performance measures shown in Table 3(b) are calculated based on the classification of test samples belonging to each cluster separately (given the membership degrees generated by the proposed semi-supervised clustering algorithm). As one can notice, the local performances of the

Table 4 (a-above) Learned weights for each classifier in each cluster with SONAR data [10]. (b-below) Per-cluster accuracy obtained using SVM, *K*-NN and Naive bayes classifiers on SONAR data [10].

Cluster #	1	2	3
SVM	**0.3775**	0.3311	0.4589
K-NN	0.3488	0.3298	**0.4702**
NBayes	0.2737	**0.3391**	0.0709

Cluster #	Cluster 1	Cluster 2	Cluster 3
SVM	0.8989	0.8430	0.8799
K-NN	**0.8577**	0.8366	**0.8831**
NBayes	0.6389	**0.8466**	0.5984

individual *K*-NN depends on the cluster. K-NN classifier 2 performs better than the other learners for samples from cluster 1. Consequently, *K*-NN classifier 2 is the most relevant classifier with respect to cluster 1. Thus, the highest aggregation weight is assigned to this classifier as reported in Table 3(a). Similarly, in cluster 3, the most accurate individual classifier is *K*-NN classifier 2.

Table 5. Performance measures of the individual learners, the method in [8], and the proposed method with SONAR dataset.

	Accuracy	Precision	Recall	F-measure
KNN	82.59	75.89	87.88	81.44
SVM	85.81	79.95	88.65	84.07
Naïve Bayes	83.37	79.87	86.23	82.92
Method in [8]	86.59	86.41	85.03	85.71
Proposed method	90.87	91.26	90.10	90.67

In the following experiment, the same feature subsets defined in the previous experiment is used, but with different classifiers. In the other words, the SONAR instances in each features subset using *K*-NN, Naive Bayes and SVM classifiers are classified. Then, the proposed semi-supervised local fusion algorithm clusters the training data, generates 3 categories, and learns optimal aggregation weights. This experiment is to indicate that our approach does not require specific classifiers, and can deal with various supervised learning algorithms.

In Table 4(a), we report the aggregation weights learned by our semi-supervised local fusion approach for each classifier with respect to the different clusters. The achievements of the different supervised learning techniques vary drastically depending on the context/cluster. More specifically, SVM is the most important learner with respect to cluster 1. This can be explained by the highest weight assigned for SVM classifier within cluster 1. Similarly, K-NN is the most relevant classifier for cluster 3.

Table 4(b) illustrates the per-cluster accuracy values obtained within the different clusters generated by our semi-supervised algorithm. As it turned out, the reported values are consistent with the relevance weights in Table 4(a). For instance, SVM which obtained the highest aggregation weight with respect to cluster 1, yields the highest accuracy with respect to this cluster. Similarly, NBayes and K-NN are the most accurate classifiers in cluster 2 and cluster 3, respectively.

Four performance measures obtained by the different individual learners, the method in [8], and our semi-supervised local fusion approach are shown in Table 5. Namely, accuracy, precision, recall and F-measure are reported for SONAR data [10]. Our approach outperforms the other methods with respect to all the performance measures.

5 Conclusion

A new approach of automatic mine detection applying SONAR dataset is proposed in this paper. This approach consists of a semi-supervised local fusion algorithm categorizing the feature space into homogeneous clusters, learning optimal aggregation weights for each classifier and optimal fusion parameters for each context in a semi supervised manner. As the experiments have shown, the semi-supervised fusion approach yields classification more accurately than the unsupervised one and the individual classifiers on synthetic and real datasets as well.

Although the proposed approach yields promising results, there will be a lot of work to do for improvement. Future works include extending the proposed approach so it deals with multiple class (more than two classes) categorization problems. Ultimately, in order to meet the need to determine the number of clusters/contexts apriori, we can study the ability of the probabilistic logic to generate duplicated clusters in order to find the optimal number of clusters.

References

1. Peyvandi, H., Farrokhrooz, M., Roufarshbaf, H., Park, S.-J.: SONAR systems and underwater signal processing: classic and modern approaches. In: Kolev, N.Z. (ed.) SONAR systems, pp. 173–206. InTech, Hampshire (2011)
2. Walsh, N.E., Walsh, W.S.: Rehabilitation of landmine victims: the ultimate challenge
3. Zamora, G.: Detecting Land Mine, November 2014. http://www.nmt.edu/mainpage/news/landmine.html
4. Miles, D.: Confronting the Land Mine Threat, November 2014. http://www.dtic.mil/afps/news/9806192.html
5. Dye, D.: High frequency sonar components of normal and hearing impaired dolphins. Master's thesis, Naval postgraduate school, Monterey, CA (2000)
6. Miles, D.: DOD Advances Countermine Technology, November 2014. http://www.dtic.mil/afps/news/9806193.html
7. Lv, C., Wang, S., Tan, M., Chen, L.: UA-MAC: an underwater acoustic channel access method for dense mobile underwater sensor networks. Int. J. Distrib. Sens. Netw. **2014**, 10 p (2014). Article ID 374028

8. Ben Ismail, M.M., Bchir, O.: Insult detection in social network comments using probabilistic based fusion approach. In: Lee, R. (ed.) Computer and Information Science. SCI, vol. 566, pp. 15–25. Springer, Heidelberg (2015)

9. Ben Ismail, M.M., Bchir, O., Emam, A.Z.: Endoscopy video summarization based on unsupervised learning and feature discrimination. In: IEEE Visual Communications and Image Processing, VCIP 2013, Malaysia (2013)

10. http://archive.ics.uci.edu/ml/datasets/Connectionist+Bench+(ines+vs.+Rocks)

11. Ben Abdallah, A.C., Frigui, H., Gader, P.D.: Adaptive local fusion with fuzzy integrals. IEEE Trans. Fuzzy Syst. **20**(5), 849–864 (2012)

12. Ben Ismail, M.M., Frigui, H.: Unsupervised clustering and feature weighting based on generalized dirichlet mixture modeling. Inf. Sci. **274**, 35–54 (2014). doi:10.1016/j.ins.2014.02.146

The Challenge of Increasing Safe Response of Antivirus Software Users

Vlasta Stavova[✉], Vashek Matyas, and Kamil Malinka

CRoCS Laboratory, Faculty of Informatics, Masaryk University,
Brno, Czech Republic
vlasta.stavova@mail.muni.cz, matyas@fi.muni.cz, malinka@ics.muni.cz

Abstract. While antivirus software is an essential part of nearly every computer, users often ignore its warnings and they are often unable to make a safe response when interacting with antivirus software. The aim of our study was to find working connections to increase a number of mobile device users who select a premium license with more security features over a free license with a limited level of security. We cooperated with the antivirus company ESET and more than fourteen thousand users participated in first phase of our experiment. We tested two new types of a user dialog on the Android platform. The first user dialog was designed with a text change and the other with a new button "Ask later". As a result, we found out that the text change increased the number of premium license purchases by 66 % in the first phase of our experiment, the version with the "Ask later" button increased this number by 25 % in the same period.

1 Introduction

User security often depends on user's ability to comprehend information and warnings. Since a user is the weakest point of the security chain, it is crucial to empower him/her to make informed decisions when cooperating with security software.

Our study aimed to find working connections between user dialog design and user security behavior when using certain components of antivirus software. We have been undertaking experiment in cooperation with a company developing antivirus software, ESET. Cooperation with the company brings us a benefit of real life experiment participants. Unlike many other studies [1,2] whose results were based on participants recruited among students or Amazon Mechanical Turk users, our study is based on real product users.

Our team consists of experts from three faculties of Masaryk University. People from Faculty of Informatics, Faculty of Social Studies and Faculty of Law have been involved. This innovative connection brings a multidisciplinary view into the experiment.

Our challenge is to increase overall user security by empowering the user to make a qualified decision on the use of antivirus software on the Android platform. Thus, we designed an experiment where we made changes in the user

© Springer International Publishing Switzerland 2016
J. Kofroň and T. Vojnar (Eds.): MEMICS 2015, LNCS 9548, pp. 133–143, 2016.
DOI: 10.1007/978-3-319-29817-7_12

dialog offering the upgrade to the one-year premium license after a trial version has expired. The effect is measured by monitoring a conversion rate of the product. The conversion rate is defined as a percentage of customers who opted for the one-year premium license out of all users. Despite all effort provided by the company so far, the conversion rate on the Android platform is still low. Our challenge is to increase this number by changes in the user dialog offering the upgrade to the one-year premium license after a trial version has expired. The second chapter is focused on related work in visual warning design and persuasive approach. The third chapter describes our experiment design. The fourth chapter concludes with experiment results and observations.

2   Related Work

User dialogs and warnings design has its place in the field of security. Despite an increasing trend of automatic decisions, there are still many problems that must be decided by a user himself. Since the user sees dozens of warnings and user dialogs every day, a general blindness to them is widely observed simply due to a process of habituation [3].

There is a common term "safe response" used for a choice that brings security benefits to the user [4]. A user dialog is considered to be successful when the safe response was selected by the user. There is a question that has been asked for many times. How to empower the user to select the safe response?

2.1   Best Practises in Visual Warning Design

An effective warning structure consists of a signal word to attract attention, identification of the hazard, explanation of consequences and directives for avoiding the hazard [5]. The other approach prefers a different structure. A good warning should contain a signal word panel with signal words, color coding and an alert symbol [6]. Since the structure is not enough to increase the power of warning, use of attractors is recommended.

Attractors are parts of warnings or user dialogs serving to attract user attention [7]. Wogalter [8] recommended to add a bold type in contrast with a standard type or to add a color in contrast with a background. Especially red and yellow are very good in increasing readability [9]. Pictorial symbols in contrast with rest of background, special effects, frames, personalization and dynamic elements also work as good attractors.

Some user dialog designs become successful, but with a great loss of usability. For example, authors in the study [7] proved that the user dialog with the greatest influence requires rewriting the most important word of the whole user dialog by a user himself. Since text rewriting makes the whole process very slow, this approach is not recommended to be widely used. Other good user dialog designs highlight important text of the warning and make the user to swipe it with his mouse or simply add 10 second delay before a decision can be made. All these features inhibit the user and empower him to comprehend the text better.

Providing an explanation is a tricky question. The study [7] proved that a detailed explanation serves as a bad attractor, but other authors [10] pointed out that a warning with a "purpose string" has a higher impact on a user over the warning without any purpose. Surprisingly, an effect of different content in a purpose string is statistically insignificant. When a hazard is communicated in an explanation, the description should be specific, complete and the most important risks should come first [8].

Text structure also influences warning effectiveness. Many studies have shown that warnings in bullets or in an outline form are considered more readable than a continuous text [11]. A common fact is that people are not reading the texts, they are scanning them. Rules following from this observation are: putting most important content first, avoid being vague, get to the point quickly and structure the text [12]. Eye tracking studies proved that the area where users really read has the F-shaped pattern [13]. They read first one or two paragraphs at the top of the text and then briefly scan down in the nearly similar shape that the letter F has.

2.2 Persuasive Approach

Apart from visual principles, a persuasive approach is also involved in our study. Persuasion can be defined as a set of influence strategies based on inner human reactions and needs. Cialdini [14] introduces six basic principles of persuasion. These principles are: Reciprocity, Commitment, Social Proof, Liking, Scarcity and Authority.

- **Reciprocity** says that people tend to reciprocate behavior towards them.
- **Commitment** speaks about fact that people like being consistent in their opinions and decisions. People who did a favor for something in past, tend to do same favor in future because they feel obliged to do so.
- **Social proof** principle is simply declaring "safety is in numbers". People in an ambiguous situation tend to behave similarly as the majority.
- **Liking** says that we are more influenced by people who are similar to us. For example, they like same things as we do.
- **Scarcity** says that rare objects are more desired by people than the widely available ones.
- **Authority** emphasizes that we are easily persuaded by people who speak to us from the position of authority.

The decoy effect is also involved in persuasive approach. It describes a change in user preferences after an introduction of a decoy option. When a user decides between two equally selected options (if presented on their own) and the decoy option is introduced, consequently one option looks more favorable and the user tends to prefer it over the other. Dan Ariely in his book [15] describes an experiment to illustrate the decoy effect. The study was conducted on MIT students. They should have selected the most favorable offer of a newspaper subscription. The first offer was to buy the online newspaper subscription for $59.

The second offer was to buy the newspaper subscription in a paper version for $125. The third offer was to buy both paper and online newspaper subscription for $125. The middle offer ($125 for the paper version) then seems without sense, because it is very unfavorable for a customer, but it has large impact on a user decision strategy. It serves as a decoy offer.

When respondents were selecting from the first and third offer only, they preferred the first offer (68 picked the first offer and only 32 the third offer) mostly. When the experiment settings changed, the decoy offer was introduced, and respondents were selecting from first, second and third offer, they preferred mostly the third one (more than 80 out of 100 picked that offer). We can observe that adding the decoy option changed the user's decision and influence him to pick a different offer.

The book [15] also mentions the power of the word "free". When something can be obtained without money, it is far more attractive than when it costs $1 or any similar small price. Word "free" works as a very powerful attractor.

People do not like making decisions and also prefer to make changeable decisions over the unchangeable ones. They do not want to lose any possibility [16].

3   Experiment Design Decisions

Our experiment was divided into two phases. First, initial phase started in December 2014, and was stopped on the first of April, and the inflow of results slowly came to an end by early May. A zero variant together with first and second variants were tested. The follow-up phase started in May 2015. Based on results from the first phase, where the most successful variant was that with a text change, we applied this text change to all variants tested in the follow-up phase. Moreover, a questionnaire about a smart phone use was included. Initial phase participants were English-speaking antivirus users mostly from USA and UK. The follow-up phase was designed in four language versions – English, German, Czech and Slovak. Results of the follow-up phase will be available at the end of 2015. We focused on the product's user dialog that appears after a trial license expired. Unfortunately, we can not influence several other factors, for example marketing campaigns running in different countries differently or users' satisfaction with the product. Similarly, we can not influence the overall product workflow – there are several ways to buy the premium license and several ways to reach this user dialog.

Experiment Limitations

Unfortunately, we could not follow several good principles that have been already introduced due to several limitations that follow from cooperation with the company.

Limitations reflecting company specific requirements must be taken into account. Only minor changes could be done in a GUI because a complete redesign was ruled out by the company. We also can not influence the whole workflow or anything out of the scope of the user dialog. Several variants can not be tested due to system limitation because implementation in the system would be costly or impossible. Some variants were canceled due to unexpected turn of events. For example, to increase attractiveness of buying the premium license, we used the principle of Reciprocity and designed a user dialog offering "something more" in addition to the user who bought the premium license, in our case it was a charity donation. Due to excessive bureaucracy connected with the donation, company ruled out this variant.

Principles Used in Design

We also made a descriptive text redesign to increase its readability and comprehension. We have used several mentioned visual design principles. As attractors we used only those that do not influence overall usability of the system. Large attractors were ruled out by the company because flashing, framing or aggressive colors do not fit the company visual style. So as an attractor we used the bold type that stresses important information which should not be overlooked by user.

As for the persuasion principle use, we added a decoy option that should give special preference to buying the premium license over using free version. The decoy option pricing was set after a negotiation with the company. We also have used principle of postponing the decision by implementing the button "Ask later". Principle of reciprocity to invoke a feeling of an obligation was used in a last variant.

Variants in Consideration

The initial screen (Fig. 1) contains a descriptive text, an offer to buy an one-year license, description of the one-year license features and action buttons. The descriptive text was: "*You can continue using the app for free. To enjoy an added level of security, purchase a license and get access to these premium features:*" The redesigned text is: "*To continue with highest level of security, purchase your license and get access to these premium features:*" We redesigned the text to make it shorter and better understandable for a user. The word "free" was removed because it stresses an undesired option of not buying the premium license.

The features description was also redesigned to be more concrete, because users with lack of technical skills may have difficulties to understand what general features description represents. Thus we pinpoint illustrative subset of functionality for each feature. For example, instead of *Take advantage of the proactive Anti-Theft at my eset.com* we recommended *Locate your missing device at my eset.com.*

Fig. 1. The initial user dialog encouraging a user to buy the one-year license.

Initial Experiment Variants

- **Var. 0** is an initial variant with no change.
- **Var. 1** uses the new redesigned text instead of the old one.
- **Var. 2** implements a button "Ask later" due to an assumption that some people do not like quick decisions and may want to make an installation later. The text remained the same. There are three buttons on the screen – "Buy", "No, thanks" and "Ask later". When a user presses "Ask later", the screen appears again after a couple of days. Button "Ask later" can be pressed 3 times at most. After the third "Ask later" pressing, the screen never appears again.

Following variants are currently involved in the follow-up phase. All variants contain also the text change taken from the first variant.

- **Var. 3** uses a principle of adding the decoy option next to the standard one. In this option a basic version is for free, a three-month license for $4.99 (the decoy option) and a one-year license for $9.99.
- **Var. 4** uses the same principle as in the first variant. In this option a basic version is for free, a one-year license for $9.99 (the decoy option) and a two-year license for $14.99.

- **Var. 5** introduces a principle of reciprocity and experiments with a business model "Pay what you can" where the user can select among three prices for the same antivirus product. Users have used an antivirus trial version for free and we assume that they may feel "obliged" to the company and buy a license. The user is asked to value his/her security and user can select a price he/she wants to pay for the product out of the three offers ($6.99, $9.99, $12.99).

Technical Solution of Initial Experiment: Only English speaking customers were involved in the study, an estimated number of respondents was 500 users per the variant. Finally, we got 14,142 participants in total after three months. Following attributes were logged in company's systems. Unfortunately, we had no other information source (for example user questionnaire) to gain more information about product users in this phase.

- Variant of the screen displayed.
- User's country.
- Summed time spent on "Premium expired screen".
- User tapped "Buy" button (yes/no).
- User actually bought the license (yes/no).
- Final decision (yes/no).
- Number of "Ask later" decisions (if applicable).
- Date – screen displayed for the first time.
- Date – user bought the license.
- Date – user tapped "No, thanks".
- Device manufacturer.
- Device resolution.
- Device model.
- Android version.

4   Results and Observations

The initial experiment ran from December 2014 to early May 2015. However, there was a marketing campaign in early March 2015. Our analyses of the data showed that this campaign had a significant impact, the trends observed from the data from first three months of experiment (December 2014 to early March 2015) are completely different from trends observed afterwards. We are currently (September 2015) investigating details of this marketing campaign, but we did not come to a rational explanation and conclusion of the causes and consequences in detail. The zero variant together with the first and the second variant were tested in the first phase. Results are described in Table 2. Participants were English speaking users of trial antivirus software running on the Android platform. All variants were randomly distributed among countries, manufacturers and device users to gain an equal representation. There were 14,142 participants in total. Half of them came from USA (49.1 %). Others came mostly from UK (33.1 %) or India (5.9 %). Nearly 90 % of them use antivirus in their mobile phones, only 10 % in tablets. As for device manufacturers, nearly half of them use

Table 1. Crosstable of results at the end of December 2014.

	Purchased	Not purchased	Total
Var. 0	34	1,099	1,133
Var. 1	52	1,114	1,166
Var. 2	36	960	996
Total	122	3,173	3,295

	Purchased	Not purchased
Var. 0	1.96%	98.04%
Var. 1	3.18%	96.82%
Var. 2	2.65%	97.35%

Table 2. Crosstable of results in early March 2015.

	Purchased	Not purchased	Total
Var. 0	77	4,780	4,857
Var. 1	125	4,731	4,856
Var. 2	87	4,342	4,429
Total	289	13,853	14,142

	Purchased	Not purchased
Var. 0	1.59%	98.41%
Var. 1	2.64%	97.36%
Var. 2	2%	98%

Samsung (48.8 %). The other half is split among many producers, for example Sony (7.4 %) or HTC (5.3 %).

We set up a null hypothesis claiming that there is no difference in a number of purchases among variants. An alternative hypothesis was claiming that the difference exists.

We have conducted a Pearson Chi-Square test ($\chi^2 = 12.062$, p < .05, df = 2). [17] Since the p-value is less than the significance level .05, we rejected the null hypothesis in favor of the alternative hypothesis and proved a difference in the number of purchases among variants.

We made a post-hoc analysis among variants based on arcsine transformation of each variant. At the significance level $\alpha = .05$ we have proved difference between the zero variant and the first variant. The difference between the zero variant and the second variant was not statistically significant.

We proved that a simple text change can provide a clearer presentation of security benefits to the user and lead to an increased uptake of a more advanced security solution.

Other Observations

Observing the data, we can see interesting trends in increase and decrease of obtaining the license. Comparing the first variant with the zero variant, the first

variant has 62 % increase in getting the license over the zero variant in December
(Table 1) this trend continued to early March when the increase was also about
66 %. Comparing the zero variant with the second variant, the second variant has
35 % increase in December but only 25 % increase in early March 2015. December
increase in obtaining the premium license was quite likely influenced by overall
Christmas shopping spree.

Average time spend on the screen is same for all variants. It is good news
for the company that new variants do not imply any delay for users. We can
observe that customers who bought the license spend more time on the screen
than customers who did not.

We have observed that people who bought the license via the second variant
did not use the button "Ask later" mostly. 96.2 % out of all customers who
obtained the license after being exposed to the second variant did not used
the button "Ask later". 3.5 % used the "Ask later" button once. Only one user
obtained a license after pressing "Ask later" twice. The current results indicate
(while still not being statistically significant) that postponing the decision does
not lead to purchase in a future.

We made also several observations based on the other attributes of collected
data. All are at the significance level $\alpha = 0.05$.

- There is a statistically significant difference in a number of purchases in India
  and USA ($\chi^2 = 15.86$, p < .001, df=1), and India and UK ($\chi^2 = 11.813$,
  p < .001, df=1). Users from USA and UK purchase statistically more than
  users from India.
- There is a statistically significant difference between zero and first variant
  among USA users ($\chi^2 = 13.98$, p < .001, df=1), whereas UK users do not
  prefer any of variants significantly.
- Tablet users buy a license more often (statistically significant) than non-tablet
  users ($\chi^2 = 42.586$, p < .001, df=1). Average conversion rate for tablet users
  is 4.4 %, whereas for non-tablet users is 1.78 %. There are no statistically
  significant preferences in variants among tablet users, but non-tablet users
  prefer the first variant significantly more.
- Comparing manufacturers who are represented by at least 500 participants,
  the highest conversion rate was observed for users of LG (3.27 %), Samsung
  (3.06 %) and Motorola (2.92 %), whereas the lowest rate was observed for
  Huawei customers (around 0.2 %). There is also a statistically significant difference in purchases among Huawei and any of the following manufacturers:
  Sony, Samsung, Motorola, LG. We also have observed statistically significant
  preferences among zero and first variant in HTC ($\chi^2 = 7.631$, p < .005, df=1)
  and Samsung ($\chi^2 = 4.264$, p < .05, df=1).

Conclusion

Our task was to increase user security by empowering him/her to select the safe
choice and obtain the premium license that offers more security features than the
free license. We have cooperated with the antivirus company ESET and 14,142

real users of their product participated in our experiment. We have rejected the null hypothesis claiming that there is no difference in a number of purchases among variants (p < .05). When comparing the number of purchases of the same version of software with better security features description, a slight difference in presenting the features implies a 62 % (December) and 66 % (March) increase in purchases as a result of using the first variant. The difference between the zero variant and the first variant with the text change was statistically significant at the significance level α=.05. Increase in the number of purchases by implementing the button "Ask later" was about 35 % in December and 25 % in March, but not enough to be statistically significant. Based on results and observations, we decided to use a text change for all variants in the follow-up experiment.

Considering limitations of our experiment, we focused strongly on user dialogs in our study and we did not take into consideration a lot of other related issues. For example, the conversion rate on the Android platform is quite likely influenced not only by the user dialog, but also with overall satisfaction with the product and with the complex product workflow which offers many ways to buy a product.

References

1. Sunshine, J., Egelman, S., Almuhimedi, H., Atri, N., Cranor, L.F.: Crying wolf: an empirical study of ssl warning effectiveness. In: USENIX Security Symposium, pp. 399–416 (2009)
2. Sotirakopoulos, A., Hawkey, K., Beznosov, K.: On the challenges in usable security lab studies: lessons learned from replicating a study on ssl warnings. In: Proceedings of the Seventh Symposium on Usable Privacy and Security, p. 3. ACM (2011)
3. Amer, T., Maris, J.-M.B.: Signal words and signal icons in application control and information technology exception messages-hazard matching and habituation effects". J. Inf. Syst. **21**(2), 1–25 (2007)
4. Bravo-Lillo, C., Cranor, L.F., Downs, J., Komanduri, S., Sleeper, M.: Improving computer security dialogs. In: Campos, P., Graham, N., Jorge, J., Nunes, N., Palanque, P., Winckler, M. (eds.) INTERACT 2011, Part IV. LNCS, vol. 6949, pp. 18–35. Springer, Heidelberg (2011)
5. Wogalter, M.S., Desaulniers, D.R., Brelsford, J.W.: Consumer products: how are the hazards perceived? In: Proceedings of the Human Factors and Ergonomics Society Annual Meeting, vol. 31, no. 5, pp. 615–619 (1987)
6. Wogalter, M.: Handbook of Warnings. Human Factors and Ergonomics. Taylor & Francis, London (2006)
7. Bravo-Lillo, C., Komanduri, S., Cranor, L.F., Reeder, R.W., Sleeper, M., Downs, J., Schechter, S.: Your attention please: designing security-decision uis to make genuine risks harder to ignore. In: Proceedings of the Ninth Symposium on Usable Privacy and Security, SOUPS 2013, pp. 6:1–6:12. ACM, New York (2013)
8. Wogalter, M.S., Conzola, V.C., Smith-Jackson, T.L.: Research-based guidelines for warning design and evaluation. Appl. Ergon. **33**, 219–230 (2002)
9. Kline, P.B., Braun, C.C., Peterson, N., Silver, N.C.: The impact of color on warnings research. In: Proceedings of the Human Factors and Ergonomics Society Annual Meeting, vol. 37, no. 14, pp. 940–944 (1993)

10. Tan, J., Nguyen, K., Theodorides, M., Negrón-Arroyo, H., Thompson, C., Egelman, S., Wagner, D.: The effect of developer-specified explanations for permission requests on smartphone user behavior. In: Proceedings of the SIGCHI Conference on Human Factors in Computing Systems, CHI 2014, pp. 91–100. ACM, New York (2014)
11. Wiebe, E.N., Shaver, E.F., Wogalter, M.S.: People's beliefs about the internet: surveying the positive and negative aspects. In: Proceedings of the Human Factors and Ergonomics Society Annual Meeting vol. 45, no. 15, pp. 1186–1190 (2001)
12. Nielsen, J.: How users read on the web (1997). http://www.nngroup.com/articles/how-users-read-on-the-web/. Accessed 1 December 2014
13. Nielsen, J.: F-shaped pattern for reading web content (2006). http://www.nngroup.com/articles/f-shaped-pattern-reading-web-content/. Accessed 1 December 2014
14. Cialdini, R.: Influence: The Psychology of Persuasion. HarperCollins, New York (2009)
15. Ariely, D.: Predictably Irrational, Revised and Expanded Edition: The Hidden Forces That Shape Our Decisions. Harper Perennial/Harper Collins, New York (2010)
16. Gilbert, D.T., Ebert, J.E.: Decisions and revisions: the affective forecasting of changeable outcomes. J. Pers. Soc. Psychol. $82(4)$, 503 (2002)
17. Corder, G., Foreman, D.: Nonparametric Statistics: A Step-by-Step Approach. Wiley, Hoboken (2014)

Weak Memory Models as **LLVM-to-LLVM** Transformations

Vladimír Štill[(✉)], Petr Ročkai[(✉)], and Jiří Barnat[(✉)]

Faculty of Informatics, Masaryk University Brno, Brno, Czech Republic
{xstill,xrockai,barnat}@fi.muni.cz

Abstract. Data races are among the most difficult software bugs to discover. They arise from multiple threads accessing the same memory location, a situation which is often hard to discern from source code alone. Detection of such bugs is further complicated by individual CPUs' use of relaxed memory models. As a matter of fact, proving absence of data races is a typical task for automated formal verification. In this paper, we present a new approach for verification of multi-threaded C and C++ programs under weakened memory models (using store buffer emulation), using an unmodified model checker that assumes Sequential Consistency. In our workflow, a C or C++ program is translated into **LLVM** bitcode, which is then automatically extended with store buffer emulation. After this transformation, the extended **LLVM** bitcode is model-checked against safety and/or liveness properties with our explicit-state model checker DIVINE.

1 Introduction

Finding concurrency-related errors, such as deadlocks, livelocks and data races and their various consequences, is extremely hard – the standard testing approach does not allow the user to control the precise timing of interleaved operations. As a result, some concurrency bugs that occur under a specific interleaving of threads may remain undetected even after a substantial period of testing. To remedy this weakness of testing, formal verification methods, explicit-state model checking in particular, can be of extreme help.

Concurrent access to shared memory locations is subject to the so called memory model of the specific CPU in use. Generally speaking, in relaxed memory models, the visibility of an update to a shared memory variable may be postponed or even reordered with other updates to different memory locations. Unfortunately, most programming and modelling languages were designed to merely mimic the principles of the underlying sequential computation machine, and therefore lack the syntactic and semantic constructs required to express

This work has been partially supported by the Czech Science Foundation Grant no. 15-08772S.

P. Ročkai—The contribution of Petr Ročkai has been partially supported by Red Hat, Inc.

J. Kofroň and T. Vojnar (Eds.): MEMICS 2015, LNCS 9548, pp. 144–155, 2016.
DOI: 10.1007/978-3-319-29817-7_13

low-level details of the concurrent computation and the memory model of the underlying hardware architecture in particular. Moreover, for obvious reasons, programmers design parallel algorithms with the *Sequential Consistency* [14] memory model in mind, under which any write to or read from a shared variable is instantaneous and immediately visible to all concurrent threads or processes – an assumption that is far from the reality of contemporary processors.

To protect from inconsistencies due to the reordered or delayed memory writes in the relaxed memory model architectures, specific low-level hardware mechanisms, such as memory barriers, have to be used. A memory barrier makes sure that all the changes done prior the barrier instruction are visible to all other processes before any other instruction *after* the barrier is executed. For more details on how memory barriers work we kindly refer the reader to technical literature. Naturally, the implementation details of a specific relaxed memory model depend on the brand and model of a CPU in use [19].

As a result, programs written in programming languages such as C do not contain enough information for the compiler to emit the code whose behaviour is both correct with respect to the incomplete specification given by the source code and at the same time as efficient as possible. A widely accepted compromise is that sequential code is guaranteed to be semantically correct, but any concurrent data access is the responsibility of the programmer. Such access needs to be guarded with various programming and modelling language addons such as builtin compiler functions, operating system calls, atomic variables with (optional) explicit memory ordering specification, or other non-language mechanisms. Since the correctness of behaviour depends on a human decision, often the resulting binary code does not do exactly what it was intended to do by its developer.

This is exactly where formal verification by model checking can help. The model checking procedure [7] systematically explores all configurations (states) of a program under analysis to discover any erroneous or unwanted behaviour of the program. The procedure can easily reveal states of the program that are only reachable under a very specific thread interleaving; clearly, such states may be very hard to reach with testing alone. Examples of explicit-state model checkers include SPIN [10], DIVINE [4], or LTSmin [12]. Unfortunately, none of the mentioned model checkers have direct support for model checking programs under relaxed memory models. Instead, should a user be interested in verification of a program under relaxed memory model, the program has to be manually (or semi-manually) augmented to capture relaxed memory behaviour.

The main contribution of our paper is in a new strategy to automate model checking of C and C++ programs under relaxed memory model without the need of modification of the interpreter used by the model checker itself. Note that interpreting C and C++ alone is a challenging task and any extension of the interpreter towards relaxed memory models would only make it harder. In fact model checkers do not typically rely on direct interpretation of C or C++ code, but use some other, syntactically simpler, representation of the original program. The model checker DIVINE, for example, interprets LLVM bitcode, which is an intermediate representation of the program created by an LLVM-based compiler.

In order to perform verification of C and C++ programs under relaxed memory model, we suggest to augment the original program and extend it with further data structures (store buffers and a cleanup thread) to simulate the behaviour of the original program under relaxed memory model. However, for the same reasons as above, we avoid direct transformation of C or C++ programs – it would require to parse the complex syntax of a high-level programming language. Instead, we apply the transformation at the level of LLVM bitcode, after the original program is translated by a C++ compiler, but before the representation is passed to the model checker for verification. This scenario allows us to completely separate the weak memory extension from the use of a model checker, hence, it allows us to use any model checker capable of processing LLVM bitcode under Sequential Consistency. Our LLVM bitcode to LLVM bitcode transformation adds store buffer data emulation to under-approximate Total Store Order (TSO) – a particular theoretical model of a relaxed memory model. The transformation is implemented within the tool called LART (LLVM Abstraction and Refinement Tool, Sect. 7.1 in [22]) that is distributed as a part of DIVINE model checker bundle, under the 2-clause BSD licence.

The rest of the paper is organised as follows. Sect. 2 lists the most relevant related work, Sect. 3 gives all the details of the LLVM transformation, Sect. 4 describes some relevant but rather technical implementation details, Sect. 5 gives details on an experimental evaluation of our approach, and finally Sect. 6 concludes the paper.

2   Related Work

The idea of using model checkers to verify programs under relaxed memory models has been discussed first in connection with the explicit-state model checker Murφ [8]. The tool was used to generate all possible outcomes of small, assembly language, multiprocessor programs using a given memory model [21]. This was achieved by encoding the memory model and program under analysis in the Murφ description language, which is an idea applied in many later papers, including this one.

To cope with the rather complex situation around memory models, theoretical models have been introduced to cover as many instances of different relaxed memory behaviours as possible. The currently most used theoretical models are the *Total Store Order* (TSO) [25], *Partial Store Order* (PSO) [25] and *x86-TSO* which is a Total Store Order enriched with interlocking instructions [16]. In those theoretical models, an update may be deferred for an infinite amount of time. Therefore, even a finite state program that is instrumented with a possibly infinite delay of an update may exhibit an infinite state space. It has been proven that for such an instrumented program, the problem of reachability of a particular system configuration is decidable, but the problem of repeated reachability of a given system configuration is not [2].

A particular technique that incorporates TSO-style store buffers into the model and uses finite automata to represent the possibly infinite set of possible contents of these buffers has been introduced in [16]. Since the state space explosion problem is even worse with TSO buffers incorporated into the model, authors of [16] extended their approach with a partial-order reduction technique later on [17].

A different approach has been taken in [11], where the algorithm to be analysed was transformed into a form where the statements of the algorithm could be reordered according to a particular weak memory ordering. The transformed algorithm was then analysed using a model-checking tool, SPIN in that case.

A lot of research has been conducted to actually detect deviation of an execution of the program on a relaxed memory model architecture from an execution under Sequential Consistency (SC). An SC deviation run-time monitor using operational semantics [18] of TSO and PSO was introduced in [6], where authors considered a concrete, sequentially consistent execution of the program, and simulated it on the operational model of TSO and PSO by buffering stores, as long as they generated the same trace as the SC execution. Another approach to detect discrepancies between a sequential consistency execution and real executions relied on axiomatic definition of memory models and (SAT-based) bounded model checking [5].

The problem of relaxed memory model computation has been addressed also in the program analysis community. Given a finite-state program, a safety specification and a description of the memory model, the framework introduced in [20] computes a set of ordering constraints that guarantee the correctness of the program under the memory model. The computed constraints are maximally permissive: removing any constraint from the solution would permit an execution that violates the specification. To address the undecidability of the problem, an abstraction from precise memory models has been considered by the BLENDER tool [13]. The tool employs abstract interpretation to deliver an effective verification procedure for programs running under relaxed memory models.

Another program analysis tool, called OFFENCE, was introduced to ensure program stability [1] by inserting a memory barrier instruction where needed – an approach also used in [17]. The problem of relaxed memory model and correct placement of synchronisation primitives is also relevant for the compiler community [9].

The problem of LTL model checking for an under-approximated TSO memory model using store buffers was also evaluated in [3], where authors proposed transformation of the DVE modelling language programs to simulate TSO.

3   Emulation of Relaxed Memory in **LLVM** Bitcode

We have chosen to provide an under-approximation of the TSO memory model, both for its simplicity and the fact that it closely resembles the memory model used by x86 computers. In this memory model, all stores are required to become visible in the same order as they are executed; however, loads can be executed before independent stores. This situation can be emulated by per-thread store buffers – stores are performed into store buffers and later flushed into main memory. Loads then have to first consult their thread's respective store buffer, and if it does not contain the address in question, proceed by consulting the main memory. Loads do not see changes that are recorded only in store buffers of other threads. We can see an illustration of the TSO memory model, and its simulation using store buffers, in Fig. 1. While in the sequentially consistent case, the result $x = 0, y = 0$ would not be possible, under TSO it is a valid output of the program, and indeed it can be proved reachable by running DIVINE on the transformed code. Note that store buffers are flushed non-deterministically, using a dedicated thread; in particular, we run a dedicated flushing thread for each worker thread.

```
     int x = 0, y = 0;

1    void thread0() {                    1    void thread1() {
2        y = 1;                          2        x = 1;
3        cout << "x = " << x << endl;    3        cout << "y = " << y << endl;
4    }                                   4    }
```

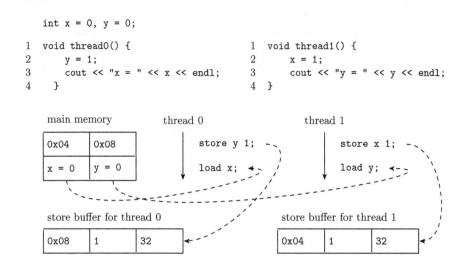

Fig. 1. In this example, each of the threads first writes into a global variable and later reads the variable written by the other thread. Under sequential consistency, the possible outcomes would be $x = 1, y = 1$; $x = 1, y = 0$; and $x = 0, y = 1$, since at least one write must proceed before the first read proceeds. However, under TSO $x = 0, y = 0$ is also possible: this corresponds to the reordering of the load on line 3 before the independent store on line 2, and can be simulated by performing the store on line 2 into a store buffer. The diagram shows (shortened) execution of the listed code. Dashed lines represent where given value is read from/stored to.

Note that we deliberately avoid precise (unbounded store-buffer) simulation of the theoretical TSO memory model, as this could easily result in infinite state space of the program under verification. However, the store buffer size can be passed as a parameter to the bitcode transformation. This way, we can make both reachability and LTL verification decidable and connect it seamlessly to an existing explicit-state framework. Please note that this approach only under-approximates the set of all TSO behaviours. I.e., when DIVINE finds a counterexample in the modified model, this counterexample can indeed occur in some runs of the given program on some real hardware with TSO semantics. On the other hand, not finding a counterexample does not guarantee error free execution on machines with store buffers deeper than specified for verification. Obviously, setting the size of store buffers is a matter of compromise – larger buffers will result in more precise verification, but also in a larger state spaces.

3.1 Infinite Delay Problem

For safety properties, such as assertion violation and/or memory safety, delaying writes indefinitely (never flushing them from a store buffer) is not a problem, as any violation of safety property is witnessed by finite path and for each run with infinite delay, there also exists (possibly finite) run where each write is eventually flushed. In infinite runs, however, such as those constructed as counterexamples to liveness properties, infinite delays could pose a problem. Imagine, for example, the following two threads:

```
bool x = false, y = false;
```

```
1  void thread0() {                1  void thread1() {
2      y = true;                    2      x = true;
3      while ( !x ) { AP( w0 ) }    3      while ( !y ) { AP( w1 ) }
4      for (;;) { /* work */ }      4      for (;;) { /* work */ }
5  }                                5  }
```

and a liveness property written (using LTL) as $FG(\neg w_0 \wedge \neg w_1)$. Assuming a separate thread to perform store buffer flushes, it is easy to see that this property holds only if the buffers are actually flushed on every possible run. However, since flushing happens non-deterministically, it may actually never happen on an infinite run. While this can be viewed as theoretically correct, it does not correspond to any real-world behaviour, where delayed writes will eventually finish and the program eventually proceeds. To counteract this inconsistency, we ask our model checker to assume weak fairness [15], where it is guaranteed that every non-blocking thread has performed infinitely many actions in an infinite run.

In [3], authors proposed to handle this problem by extending LTL specification to include this store buffer fairness criteria. In our case though, we have chosen to implement our transformation in a way which does not require any additional specification and store buffer fairness is implied by the standard weak fairness.

3.2 Invalidated Variable Store Problem

Another issue to deal with are delayed flushes from a store buffer that come at the time when the object that should be written into does not exist anymore in the main memory. As both memory allocation and stack depth can change at the run-time, it might happen that an entry in the store buffer points to invalid location (either given memory chunk was deallocated by the user, or it lived in a stack frame that has already been abandoned). To solve this problem, we would need to make sure that inaccessible addresses are evicted from the store buffers. For dynamic memory, this can be done by overriding the function which deallocates objects from memory in such a way that it first iterates over all store buffers and evict entries into the to-be-freed memory before calling the original deallocate function.

For stack memory, however, the situation is more complicated – it is not sufficient to evict all the stack-frame-allocated memory from store buffers before returning from a function, because an exception can cause stack unwinding, which can also result in invalid references in store buffers. This means that cleanup handlers [24] need to be added to each function to deal with the situation.

4   Implementation

First of all, let us briefly explain how LLVM bitcode is used by our target model checker DIVINE to support for C/C++ verification. There are two levels below the LLVM bitcode of the program to be verified – an interpreter and an LLVM *userspace*. The interpreter is used directly by the model checker to generate and explore the state space graph by executing LLVM instructions. The interpreter detects errors such as invalid memory dereference, memory leaks, assertion violations, etc. The interpreter has to be aware of threads and dynamic memory management, hence, its role is similar to what the CPU and the core of the operating system do when executing the code natively. The userspace, on the other hand, corresponds to the runtime of the programming language, that is, it provides LLVM bitcode for the basic libraries required by the given programming language and/or threading model. The userpsace and interpreter together provide the user with a standards-compliant interface for user's programming language of choice.

While in general, the separation of work between the interpreter and userspace could be almost arbitrary (one could, for example, include the entire pthread library in the interpreter), it is advantageous to keep the interpreter as simple as possible, pushing most of the required functionality into the userspace. Therefore, DIVINE provides a fairly small set of intrinsic functions (sixteen in total), which give access to the necessary functionality provided by the interpreter. The rest is left to userspace.

The support for relaxed memory verification, such as functions that simulate store buffers, thus need not come separately for every program to be verified under relaxed memory model, but may actually become a part of the DIVINE LLVM userspace. However, it is not possible to implement weak memory simulation through addition of userspace functions alone – we need to change the behaviour of memory manipulation instructions (such as loads, stores, and fences). For this reason, we implemented an LLVM to LLVM bitcode transformation pass, which translates relevant instructions into calls to the relevant userspace functions. The actual simulation of the memory model is thus implemented within the userspace and is separate from the original program. As a result of this design choice, this transformation can be easily modified to work with other LLVM model checkers and with different weak memory models.

4.1 Updates to **LLVM** Userspace

Currently, LLVM userspace provides replacement functions for `load`, `store` and `fence`. The relevant userspace functions can be identified by their `__lart_weakmem` prefix. Store buffers are represented by a thread-local array with one record for each store – this record contains the address, the value itself and the bit width of the value. We have chosen to limit a single store to 64 bits, which is the usual size atomically written by modern CPUs and also the maximal size of standard integer types in C. Each store then pushes a record into the local store buffer, while loads first consult the local store buffer for an up-to-date value, and if it is not present proceed to load from memory. A fence flushes all the entries from the local store buffer.

Note that block memory manipulation functions have to be replaced too, to protect them from bypassing the store buffers. Hence, the userspace provides replacements for block memory manipulation functions such as `llvm.memmove`, `llvm.memcpy`, etc.

Further, atomic LLVM instructions, e.g. `cmpxchg`, are rewritten within the transformation to use only functions implemented within the userspace. However, we currently only support sequentially-consistent ordering of atomics (which is the default ordering for atomic variables in C++11). Further extensions to support all atomic access orderings supported by LLVM/C++11 are planned.

Finally, attention had to be paid to initialisation of the store buffers. Due to the nature of global variable constructors in C++ which can run in arbitrary order, we cannot use non-trivial constructors for store buffers, as this could cause the constructor to run after some calls to `__lart_weakmem_*` functions have already happened. Therefore, the store buffer array is initialised to a null pointer and allocated in the first call to one of the `__lart_weakmem_*` functions.

4.2 **LLVM** to **LLVM** Transformation

The transformation is implemented as part of the LART tool. It basically iterates over all the instructions in the original LLVM bitcode and replaces some of them with calls to the corresponding replacement functions.

To perform this transformation correctly, we had to introduced special LLVM function attributes: *bypass*, *tso*, and *sc*, denoting in what mode a particular function should operate. Functions marked *bypass* are not subject to the transformation at all, functions marked *tso* are fully processed by the transformation as indicated above. In functions marked *sc*, additional memory barriers are inserted at the beginning of the function and after a call to any non-SC function. Note that it is important that the functions which implement the relaxed weak memory model itself are not transformed; for this reason, all `__lart_weakmem_*` functions are annotated as *bypass*. The default behaviour of the transformation on functions that are not annotated with any of the attributes can be set by a parameter passed to the transformation.

Since LLVM allows loads and stores larger than 64 bits (either large scalar types, such as 128 bit integers, or aggregate values), we first break these large loads and stores into chunks of at most 64 bit-wide operations in a separate transformation pass and only after this is done, we perform the instruction substitution transformation as outlined above.

Finally, to avoid interference from compiler optimisations, some of the memory accesses in our functions had to be marked volatile and we had to prevent inlining of some of the functions (since inlining would discard function attributes). Likewise, all the exposed functions had to be marked noinline.

4.3   State Space Reduction

Store buffers substantially increase the size of the state space, hence it is necessary to counteract this growth. DIVINE provides powerful reduction techniques out of the box, based on analysis of instruction visibility. Those reductions are, however, rendered less effective by interactions with the store buffer: in particular, any TSO load or store is treated as visible by the $\tau+$ reduction due to global variable access within the TSO load/store implementation.

Fortunately, it is possible to reduce the overhead of store buffers by entirely bypassing their use for memory locations that are private to a particular thread. However, since the entire logic of TSO stores is handled in the userspace, it is necessary to expose an additional intrinsic (builtin) function in the model checker, which, for a given address, decides whether the address is visible from any other threads.

As far as correctness is concerned, when we realise that from the point of view of the model checker, store buffers are part of the global memory, the argument carries over from the analogical construct (store visibility) used in $\tau+$ reduction [23]. Any pointers currently residing in store buffers – and hence, capable of revealing new memory locations to foreign threads – are treated as global; hence, a delayed write of such a pointer cannot incorrectly hide intervening stores (into locations that were previously thread-private but revealed by the pointer living in a store buffer).

5   Evaluation

We evaluated our approach on a few models, all of which can be found in examples in source distribution of DIVINE[1]. Descriptions of the models used can be found in Table 1. All measurements were performed on a laptop with Intel Core i7-3520M, running at 3.4 GHz, with 8 GB of memory. DIVINE used 4 threads for verification and never depleted available memory (loss-less state space compression was enabled).

5.1   Results

The results of verification with DIVINE can be seen in Table 2. In all cases, Context-Switch-Directed Reachability [26] was used, as it performed much faster

---

[1] Online: https://divine.fi.muni.cz/trac/browser/examples/llvm/weakmem/.

Table 1. Models used for evaluation

`simple_sc`	Model based on Fig. 1, SC, asserting that $x = 0, y = 0$
`simple_mtso`	Same model, but manually modified to use TSO for relevant variables
`simple_stso`	Same model, workers are auto-transformed to TSO, the rest is SC
`simple_tso`	Same model, fully transformed to TSO
`peterson_sc`	Peterson's mutual exclusion algorithm
`peterson_tso`	The same, automatically transformed to TSO.
`fifo_sc`	First-in, first-out, lockless inter-thread queue, as used in DIVINE
`fifo_tso`	Automated TSO transform of `fifo_sc` above

Table 2. Results of `divine verify` for our examples.

Model	Store buffer size	Assertion violated	# of states	Reduced # states	Memory [GB]	Time [s]
`simple_sc`	N/A	no	205	N/A	0.16	1
`simple_mtso`	1	yes	6.89 k	N/A	0.17	3
`simple_stso`	1	yes	10.7 k	10.7 k	0.17	6
`simple_tso`	1	yes	24.7 M	537.2 k	3.18	20318
`peterson_sc`	N/A	no	1.68 k	N/A	0.16	1
`peterson_tso`	0	no	55.9 k	N/A	0.17	38
`peterson_tso`	2	yes	2.86 M	95.7 k	0.79	990
`peterson_tso`	3	yes	4.70 M	129.9 k	1.21	1610
`fifo_sc`	0	no	6951	N/A	0.73	20
`fifo_tso`	1	no	–	44 M	–	–

than regular reachability for the TSO simulation case. From the results, we can see significant increase of state space size when store buffers are enabled. This is due to two factors – one of them is that the store buffers themselves increase the state space size, as they can be flushed non-deterministically anywhere between the given store and the nearest memory barrier. The other issue is the interference with $\tau+$ reduction mentioned in Sect. 4.3. As can be seen in the case of `peterson_sc` and `peterson_tso` with store buffers of size 0 (in this case value is stored into store buffer and immediately flushed out within one transition in the state space), this effect is quite strong.

As for the differences between different versions of the `simple` model, the state space size is clearly dependent on how many of the loads and stores are treated as TSO – in case of full TSO transformation all library functions are also in TSO, therefore state space size is increased far more. The difference between `simple_mtso` and `simple_stso` is more subtle: in the case of `simple_stso` our transformation adds memory barriers into SC functions, at their beginning and after any call to non-SC function. While the second case is rarely present in our model, the first case makes any function call observable, as a flush will be considered observable by $\tau+$ reduction (due to an accesses to the store buffer).

6   Conclusion

We have introduced an LLVM to LLVM transformation that extends a program with relaxed memory simulation and we have shown that such an extended program can be passed to a model checker to perform verification of C/C++ programs under a relaxed memory model. A key attribute of our approach is that no updates to the model checker (which is based on sequential consistency) are needed. The preliminary experiments show the approach as such is feasible, even though the growth of the state space is significant. Finally, the verification of the `fifo_tso` model is, in itself, a valuable result, as the code in question is sensitive to memory ordering and until now we were only able to verify it under the assumption of sequential consistency.

As our future work we intend to improve the implementation and also implement support for weaker memory models, such as Partial Store Order. As a research goal, we want to extend LART to automatically annotate some functions as SC, whenever it can be statically decided that such an annotation has no influence on the verification result, counteracting the growth of the state space. Further improvements of reductions supported by DIVINE and their interaction with store buffer simulation, and thread-local memory in general, could also significantly reduce the state space.

References

1. Alglave, J., Maranget, L.: Stability in weak memory models. In: Gopalakrishnan, G., Qadeer, S. (eds.) CAV 2011. LNCS, vol. 6806, pp. 50–66. Springer, Heidelberg (2011)
2. Atig, M.F., Bouajjani, A., Burckhardt, S., Musuvathi, M.: On the verification problem for weak memory models. In: Proceedings of the 37th annual ACM SIGPLAN-SIGACT Symposium on Principles of Programming Languages, POPL 2010, pp. 7–18. ACM, New York, NY, USA (2010)
3. Barnat, J., Brim, L., Havel, V.: LTL Model checking of parallel programs with under-approximated TSO memory model. In: Application of Concurrency to System Design (ACSD), pp. 51–59. IEEE (2013)
4. Barnat, J., Brim, L., Havel, V., Havlíček, J., Kriho, J., Lenčo, M., Ročkai, P., Štill, V., et al.: DiVinE 3.0 - an explicit-state model checker for multithreaded C & C++ programs. CAV 2013. LNCS, vol. 8044, pp. 863–868. Springer, Heidelberg (2013)
5. Burckhardt, S., Musuvathi, M.: Effective program verification for relaxed memory models. In: Gupta, A., Malik, S. (eds.) CAV 2008. LNCS, vol. 5123, pp. 107–120. Springer, Heidelberg (2008)
6. Burnim, J., Sen, K., Stergiou, C.: Sound and complete monitoring of sequential consistency in relaxed memory models. In: Technical report UCB/EECS-2010-31, EECS Department, University of California, Berkeley, March 2010
7. Clarke, E., Grumberg, O., Peled, D.: Model Checking. MIT Press, Cambridge (1999)

8. Dill, D.: The murphi verification system. Computer Aided Verification. LLNC, vol. 1102, pp. 390–393. Springer, Heidelberg (1996)
9. Fang, X., Lee, J., Midkiff, S.P.: Automatic fence insertion for shared memory multiprocessing. In: International Conference on Supercomputing (ICS 2003), pp. 285–294. ACM (2003)
10. Holzmann, G.J.: The Spin Model Checker: Primer and Reference Manual. Addison-Wesley, Reading (2004)
11. Jonsson, B.: State-space exploration for concurrent algorithms under weak memory orderings: (preliminary version). SIGARCH Comput. Archit. News **36**, 65–71 (2009)
12. Kant, G., Laarman, A., Meijer, J., van de Pol, J., Blom, S., van Dijk, T.: LTSmin: high-performance language-independent model checking. In: Baier, C., Tinelli, C. (eds.) TACAS 2015. LNCS, vol. 9035, pp. 692–707. Springer, Heidelberg (2015)
13. Kuperstein, M., Vechev, M., Yahav, E.: Partial-coherence abstractions for relaxed memory models. In: Programming Language Design and Implementation (PLDI 2011), pp. 187–198. ACM (2011)
14. Lamport, L.: How to make a multiprocessor computer that correctly executes multiprocess programs. IEEE Trans. Comput. **28**(9), 690–691 (1979)
15. Lehmann, D.J., Pnueli, A., Stavi, J.: Impartiality, justice, fairness: the ethics of concurrent termination. ICALP. LNCS, vol. 115, pp. 264–277. Springer, Heidelberg (1981)
16. Linden, A., Wolper, P.: An automata-based symbolic approach for verifying programs on relaxed memory models. In: van de Pol, J., Weber, M. (eds.) Model Checking Software. LNCS, vol. 6349, pp. 212–226. Springer, Heidelberg (2010)
17. Linden, A., Wolper, P.: A verification-based approach to memory fence insertion in relaxed memory systems. In: Groce, A., Musuvathi, M. (eds.) SPIN Workshops 2011. LNCS, vol. 6823, pp. 144–160. Springer, Heidelberg (2011)
18. Mador-Haim, S., Alur, R., Martin, M.M.K.: Specifying relaxed memory models for state exploration tools. In: $(EC)^2$: Workshop on Exploting Concurrency Efficiently and Correctly (2009)
19. Mckenney, P.E.: Memory Barriers: a Hardware View for Software Hackers. Linux Technology Center, Beaverton (2009)
20. Kuperstein, M., Vechev, M.T., Yahav, E.: Automatic inference of memory fences. In: Formal Methods in Computer-Aided Design, pp. 111–119. IEEE (2010)
21. Park, S., Dill, D.: An executable specification and verifier for relaxed memory order. IEEE Trans. Comput. **48**(2), 227–235 (1999)
22. Ročkai, P.: Model checking software. Disertation thesis, Faculty of Informatics, Masaryk University (2015)
23. Ročkai, P., Barnat, J., Brim, L.: Improved state space reductions for LTL model checking of C and C++ programs. NFM 2013. LNCS, vol. 7871, pp. 1–15. Springer, Heidelberg (2013)
24. Ročkai, P., Barnat, J., Brim, L.: Model checking C++ with exceptions. In: Automated Verification of Critical Systems, vol. 70 (2014)
25. CORPORATE SPARC International, Inc.: The SPARC architecture manual (version 9). Prentice-Hall Inc., Upper Saddle River (1994)
26. Štill, V., Ročkai, P., Barnat, J.: Context-switch-directed verification in DIVINE. In: Hliněný, P., Dvořák, Z., Jaroš, J., Kofroň, J., Kořenek, J., Matula, P., Pala, K. (eds.) MEMICS 2014. LNCS, vol. 8934, pp. 135–146. Springer, Heidelberg (2014)

Author Index

Printed in the United States
By Bookmasters